# ARCHITECTURAL LIGHTING DESIGN

# ARCHITECTURAL LIGHTING DESIGN

Frederic H. Jones, Ph.D.

Crisp Publications, Inc.

Los Altos, California

**Architectural Lighting Design**
by Frederic H. Jones, Ph.D.


Library of Congress Catalog Card Number 000000000
ISBN 0-931961-93-9

All rights reserved. No part of this book may be reproduced or transmitted in any form or by any means now known or to be invented, electronic or mechanical, including photocopying, recording, or by any information storage or retrieval system without written permission from the author or publisher, except for the brief inclusion of quotations in a review.

Copyright 1989 by Frederic H. Jones, Ph.D.
Printed in the United States of America
Book design and typesetting by Frederic H. Jones
Editorial advice by Judith K. Jones
Cover design by Kennon-Kelley Graphic Design


### ACKNOWLEDGEMENTS

The author and publisher wish to thank Lightolier, Inc. for their generous permission to utilize, almost in its entirety, their *Lessons in Lighting* in this book. I have endeavored to expand on the sections I have used. My thanks also to Marlene Lee Lighting Design for allowing the use of specification and working examples in Chapters 5 and 6, and to the Illuminating Engineering Society for allowing the reprinting of a number of illustrations and charts. All illustrations are credited in the Plate List at the end of the book.

# TABLE OF CONTENTS

**1. THE DESIGN MEDIUM** — 3
THE PROCESS OF VISION — 4
PHYSIOLOGICAL FACTORS — 6
    DEFINITIONS — 8
        VISUAL FIELD — 8
        ACCOMMODATION — 8
        ADAPTATION — 8
    THE EYE AND AGE — 8
PHYSICAL FACTORS — 9
    DEFINITIONS — 9
        SIZE — 9
        CONTRAST — 9
        TIME — 10
    QUANTITY OF LIGHT — 10
        DEFINITIONS — 11
        CANDLEPOWER DISTRIBUTION CURVES — 11
    QUALITY OF LIGHT — 12
        GLARE — 12
        VISUAL COMFORT PROBABILITY (VCP) — 12
        VEILING REFLECTIONS — 14
        EQUIVALENT SPHERE ILLUMINATION (ESI) — 14
    LIGHTING TECHNOLOGY — 15
        TRANSMISSION — 21
            DIRECT TRANSMISSION — 22
            SPREAD TRANSMISSION — 22
            DIFFUSE TRANSMISSION — 22
        REFRACTION — 22
        REFLECTION — 23
            SPECULAR REFLECTION — 23
            SPREAD REFLECTION — 23
            DIFFUSE REFLECTION — 23
        CONTROL OF THE BEAM — 24
        ABSORPTION — 25
    LUMINAIRES — 25
        CLASSIFICATION OF LUMINAIRES — 25
            DIRECT — 26
            SEMI DIRECT — 27
            GENERAL DIFFUSE — 28
            DIRECT-INDIRECT — 29
            SEMI-INDIRECT — 30
            INDIRECT — 31

|                                              |    |
|----------------------------------------------|----|
| LUMINAIRE SPECIFICATIONS                     | 32 |
| PHOTOMETRIC DATA                             | 32 |
| POLAR GRAPH                                  | 32 |
| EFFICIENCY                                   | 34 |
| SPACING RATIO                                | 34 |
| NON-SYMMETRICAL DISTRIBUTION                 | 34 |
| COEFFICIENTS OF UTILIZATION                  | 36 |
| HOW TO USE PHOTOMETRIC DATA                  | 37 |
| CONTROL OF LUMINAIRE BRIGHTNESS              | 40 |
| SELECTING A LUMINAIRE                        | 41 |
| SUMMARY                                      | 42 |
| **2. DESIGN PRINCIPLES**                     | 43 |
| THE ART AND SCIENCE OF LIGHTING              | 43 |
| PRINCIPLES                                   | 44 |
| EFFECT ON ARCHITECTURE                       | 44 |
| EFFECT ON INTERIOR DESIGN                    | 44 |
| PLANES OF BRIGHTNESS                         | 45 |
| GLITTER AND SPARKLE                          | 46 |
| LIGHT AND SHADOW                             | 47 |
| MODELING                                     | 48 |
| LIGHTING DESIGN CONSIDERATIONS               | 49 |
| KINDS OF LIGHTING                            | 51 |
| TECHNIQUES                                   | 54 |
| OVERVIEW                                     | 54 |
| SURFACES                                     | 54 |
| LIGHTING THE VERTICAL SURFACE                | 54 |
| ACCENT LIGHTING                              | 58 |
| DESIGNING THE WORKSPACE                      | 60 |
| TASK VISIBILITY                              | 60 |
| LIGHTING THE TASK                            | 62 |
| GLARE                                        | 66 |
| REFLECTED GLARE                              | 66 |
| DIRECT GLARE                                 | 67 |
| LOCAL LIGHTING                               | 68 |
| OPEN PLAN TASK LIGHTING                      | 68 |
| LOCATION OF TASK LIGHTING                    | 68 |
| CONTROLLED TASK LIGHTING                     | 69 |
| SUMMARY                                      | 70 |
| **3. LIGHTING ENGINEERING**                  | 71 |
| CALCULATIONS                                 | 72 |
| DESIGN CONSIDERATIONS                        | 72 |
| POINT METHOD FOR SINGLE LUMINAIRES.          | 73 |
| HOW TO CALCULATE FOOTCANDLES AT A POINT      | 73 |
| CALCULATOR CHARTS FOR SINGLE LUMINAIRES      | 75 |

| | |
|---|---|
| ZONAL CAVITY METHOD FOR AVERAGE ILLUMINATION | 78 |
|     METHOD PROCEDURE | 79 |
|     QUICK CALCULATOR CHARTS | 85 |
| COMPUTER CALCULATIONS | 89 |
|     LUMEN-MICRO | 92 |
| SUMMARY | 96 |
| **4. LIGHTING TECHNOLOGY** | 97 |
| LIGHT SOURCES | 98 |
|     INCANDESCENCE | 98 |
|     PHOTOLUMINESCENCE | 98 |
|     SOURCE CHARACTERISTICS | 99 |
| INCANDESCENT LAMPS | 101 |
|     NOMENCLATURE | 101 |
|     BASES | 103 |
|     BULB FINISHES | 103 |
|     EFFICACY | 103 |
|     LAMP DEPRECIATION | 104 |
|     COLOR TEMPERATURE | 104 |
|     CLASSES OF LAMPS | 105 |
|     ADVANTAGES OF INCANDESCENT SOURCES | 106 |
|     DISADVANTAGES OF INCANDESCENT SOURCES. | 106 |
| GAS DISCHARGE LAMPS. | 107 |
| FLUORESCENT LAMPS | 108 |
| HIGH INTENSITY DISCHARGE LAMPS | 114 |
| BALLASTS | 120 |
|     BALLAST WATTAGE | 121 |
|     BALLAST LIFE | 121 |
|     POWER FACTOR | 121 |
|     BALLAST PROTECTION | 122 |
|     BALLAST NOISE | 122 |
|     DIMMING | 123 |
|     FLUORESCENT BALLASTS | 123 |
|     HID BALLASTS | 124 |
|     MERCURY BALLASTS | 124 |
|     METAL HALIDE BALLASTS | 125 |
|     HIGH PRESSURE SODIUM | 125 |
| SUMMARY | 126 |
| **5. LIGHTING GRAPHICS** | 127 |
| INTRODUCTION | 127 |
| **6. SPECIFICATIONS** | 145 |
| INTRODUCTION | 145 |
| SAMPLE SPECIFICATION | 145 |
| **GLOSSARY** | 167 |
| **BIBLIOGRAPHY** | 181 |
| **LIST OF PLATES** | 183 |

# Introduction

The intention of this book is not to provide the last word or even the most definitive word on architectural lighting. The goal is rather to be a portfolio of design and technical information that can serve as a springboard to more technical references and texts while providing substantial information for the architect and designer who are engaged in lighting their own designs or are working with a professional lighting designer.

Lighting is a complex and ethereal aspect of the architectural design process. Very few architects and engineers ever give more than lip service to truly designing the lighting environment. Too often, the job is shunted off to the electrical engineer or the lighting sales representative. Furthermore many have the false impression that a good-looking reflected ceiling plan is a lighting design. Most electrical engineers calculate lighting levels and provide layouts that assure the client of a maintained lighting level. Typical layouts and calculations traditionally have consisted only of these concerns, with the recent addition of energy consumption problems. While many manufacturer's representatives are quite knowledgeable, they primarily know products, not design. The architect often fails to realize that the problems of visual comfort and reflected glare can have a greater impact on the effectiveness of the lighting installation than does the pattern of fixture placement or initial luminance.

# Chapter 1

## THE DESIGN MEDIUM

There is a great deal more to the art and science of lighting than physics and technology. The process of lighting design must also take into account the relationship between the luminous environment and the eye and mind of the human observer. The psychology of the environment is, of course, always at play in the perception and relationship of people and architecture, but in the case of lighting the relationship is heightened. The actual way the eye/mind combination perceives light is a dynamic thing and bears at least equal value in the lighting design process. In this chapter we will consider both the process of vision and the effect of lighting and perception on the experience of the lighted architecture itself. You should come away with a better understanding of the importance of these considerations in both the design and implementation of the luminous architectural environment.

## THE PROCESS OF VISION

In order to see there must be light, an object, a receptor (the eye) and a decoder (the brain). LIGHT is electromagnetic energy emitted in the visible portion of the spectrum. White light results from combining different wavelengths of visible energy.

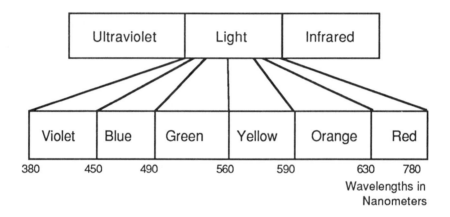

Spectrum. Fig 1-1

All the light below 380 nanometers is ultraviolet and above 780 is infrared. The eye is unable to perceive these wave lengths but the skin responds. Skin is tanned as a result of the effect of ultraviolet and is warmed by infrared. The visible spectrum contains all the colors of the rainbow and when it is seen in combination is perceived as white light.

The light output of a source is measured in LUMENS. The INTENSITY of light (luminous intensity) in a given direction is measured in CANDELAS. When light strikes a surface it is measured in FOOTCANDLES-- one footcandle (fc) being the illumination on the surface one foot away from a standard candle (or one lumen per square foot).

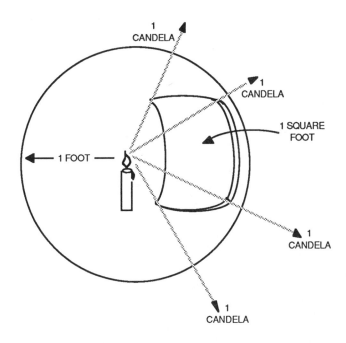

**Footcandle Chart. Fig 1-2**

The figure above represents a sphere equidistant from the point of the light source. The inside surface of the sphere is one foot away from the candle flame and one square foot of light falling on the surface represents one foot candle.

We do not see footcandles. Rather, we see the brightness resulting from light transmitted or reflected by a surface. This BRIGHTNESS is measured in FOOTLAMBERTS. There is always a subtractive interaction between a surface and the light falling on it. Light energy is absorbed or interfered with by surfaces. Some light is always lost.

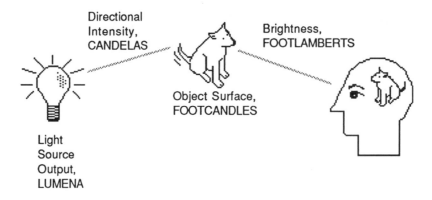

**Chart of Vision. Fig 1-3**

The light rays reflected or transmitted from the object whose brightness we see stimulate electrochemical receptors in the eye that transmit signals to the brain. The brain and the eye cooperate in transforming radiant energy into the sensation of vision.

This is a clear representation of the system nature of vision and indicative of the need to be aware of all the elements that combine to make vision possible. We will next consider the effects of human physiology on the vision process and the differences between age and youth in vision.

## PHYSIOLOGICAL FACTORS

The eye responds to electromagnetic energy-wavelengths in the range between ultraviolet and infrared radiation. The eye is most responsive to the yellow-green portion of the spectrum.

The eye is a complex mechanism that can be understood by comparison to the camera. It consists of a lens system, an image receptor and a mechanism to focus and control the various relationships. The iris and pupil are analogous to the diaphragm and shutter of the camera and the lens, of course, relates to the camera lens. The retina is similar to the camera film. There are two kinds of nerve receptors that receive and transmit light stimulus to the brain. These occur in the retina and are called rods and cones. Rods are responsible for daylight and color vision while cones specialize in dawn and dusk vision. Malfunction of the rod system can result in color blindness while damaged cones can result in night blindness. The vitreous humor refracts light and assists the lens in controlling the vision process.

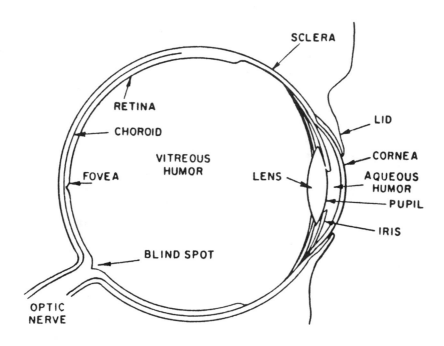

Eye Chart. Fig 1-4

## DEFINITIONS

### VISUAL FIELD

The VISUAL FIELD is the area the eye sees. This field normally extends 10 degrees in the vertical. The finest details are seen only in a small area at the back of the eye known as the fovea. Details become progressively less distinct as they approach the outer limit of the visual field although movement and changes in brightness levels remain readily discernible even at the periphery.

### ACCOMMODATION

ACCOMMODATION is the process by which the eye locates and focuses on an object. The eye physically changes shape in order to accommodate the distance from the object. The nearer the object, the more convex the lens of the eye will be. The farther away the object, the flatter the lens. Prescription glasses compensate for the inability of the lens to change shape sufficiently to provide clear vision.

### ADAPTATION

ADAPTATION involves the size of the pupil opening and the sensitivity of the retina. The pupil of the eye opens wide in low levels of light and gets smaller as the light level increases. The chart indicates the range of lighting level, from moonlight (about .01 footcandles) to tropical noon, in which humans see. A change also occurs in the photochemical substances of the retina. It takes longer to adapt from light to dark, for example going into a movie theater in the daytime, than it does to adapt from dark to light.

## THE EYE AND AGE

Twenty/twenty vision is what normal twenty year olds can see at twenty feet. The eyes of a healthy twenty year old adjust quickly and easily to changes in brightness in the environment. As eyes age, they lose their elasticity, reducing their ability to accommodate easily. Adaptation from one light level to another takes longer and the range of sensitivity drastically diminishes the ability to see at low light levels. A sixty year old needs ten time as much light as a normal twenty year old to perform the same seeing task with equal speed and accuracy.

## PHYSICAL FACTORS

### DEFINITIONS

### SIZE

The four factors that determine visibility are:

SIZE. The bigger or nearer an object, the easier it is to see.

Seeing becomes more difficult ⟶

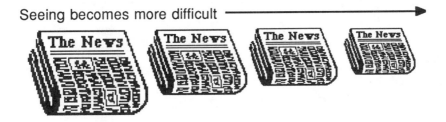

**The nearer an object, the easier to see. Fig 1-5**

### CONTRAST

CONTRAST. The difference in brightness (luminance) of an object and its background. Black letters on white paper are easy to read because the contrast approaches 100% but grey lettering with only 40% reflectance on grey paper of 80% will only have a contrast of 50% and be hard to see. Visibility can be increased by adding illumination or the use of color.

Seeing becomes more difficult ⟶

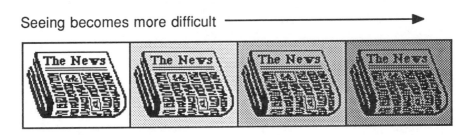

**More contrast means increased visibility. Fig 1-6**

## LUMINANCE

LUMINANCE is the proper term for what is often called brightness. Luminance is the amount of light reflected or transmitted by an object. A light colored surface reflects more light than a dark one, hence more illumination is needed on a dark surface to equal the luminance of a similar light surface.

Luminance - reflected light. Fig 1-7

## TIME

TIME or how long it takes to see. The less light, the longer it takes to see details. The time factor is especially important where motion is involved, as in driving. Under low light levels, an object appears to move more slowly than under high levels of illumination.

## QUANTITY OF LIGHT

Electric light sources, whether incandescent, fluorescent or high intensity discharge, are known in the industry as LAMPS. Except for decorative bulbs, bare lamps are a source of glare, therefore they are shielded and their output is controlled by LUMINAIRES (fixtures). Some lumens are always lost in a luminaire by being blocked or absorbed. LUMINAIRE EFFICIENCY is the ratio of the lumen output of the luminaire to the lumen output of the lamp, expressed as a percentage.

## DEFINITIONS

### INVERSE SQUARE LAW

We know that the basic unit of measurement of light is a LUMEN and light arriving on a surface is measured in footcandles. A FOOTCANDLE is the amount of illumination from one standard candle falling on a surface one foot away. As the distance between the surface and the candle (a point source) increases, the CANDELAS (candlepower) reaching that surface at a given point decrease according to the INVERSE SQUARE LAW; that is, the candlepower divided by the square of the distance equals footcandles ($cp/D^2=fc$).

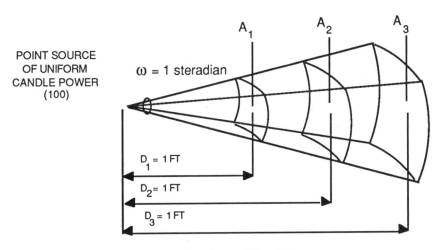

**Inverse Sq Law. Fig 1-8**

For example, 100 candelas lighting a surface one foot away provide 100 fc, but at 2' away, only 25 fc and at 3', 11 fc. Because of their geometry, illumination from linear and area sources is calculated in a different way.

### CANDLEPOWER DISTRIBUTION CURVES

CANDLEPOWER DISTRIBUTION CURVES as shown in Fig. 1-9, show graphically the intensity of luminous flux (candelas) in a given direction.

The ratio of the quantity of lumens leaving a light source in a luminaire to that which arrives on the working plane as useful light is called the COEFFICIENT OF UTILIZATION. CU TABLES indicate the combined efficiency of the luminaire, room proportions and room reflectances and are used to calculate illumination levels.

The United States is changing over to the international metric system in which LUX is the counterpart of a footcandle. To approximate lux quickly, multiply footcandles by ten. For more accurate conversion, multiply by 10.76 (1 fc = 10.76 lx).

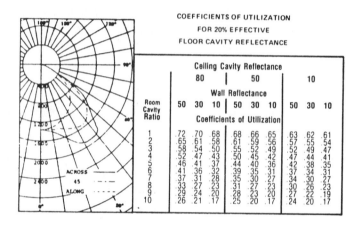

Candlepower Distribution Curves from catalog. Fig 1-9

# QUALITY OF LIGHT

## GLARE

GLARE is brightness in the visual field which is annoying and uncomfortable, causing fatigue and loss of productivity. DIRECT GLARE results from seeing high luminaire brightness in the normal field of view.

## VISUAL COMFORT PROBABILITY (VCP)

VISUAL COMFORT PROBABILITY (VCP) evaluates the probability in a given set of parameters that a person seated in the worst position for glare in a room (usually at the center rear) will find the degree of glare from the lighting system just acceptable. For example, an installation of 2' x 4' lens fixtures with a VCP of 75 would signify that 75% of the people in the worst position would find it visually acceptable and 25% would not.

Chapter One   The Design Medium   13

**Glare in the field of view causes discomfort. Fig 1-10**

The Illuminating Engineering Society recommends a minimum VCP of 70 and a ratio of maximum-to-average luminaire brightness of five to one, with a ratio of three to one being preferable. IES recommendations for maximum luminance at various angles are:

| Degrees from Nadir | Maximum Luminance (fL) |
|---|---|
| 45 | 2250 |
| 55 | 1605 |
| 65 | 1125 |
| 75 | 750 |
| 85 | 495 |

REFLECTED GLARE causes discomfort and occurs when luminaire brightness is reflected from shiny (SPECULAR) surfaces in the field of view.

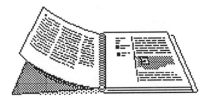
No veiling reflections

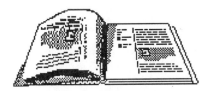
Veiling reflections

**Veiled Reflections. Fig 1-11**

VEILING REFLECTIONS are more subtle reflections of the light source in the task. They reduce the contrast between detail and background as shown in figure 1-11 and thus reduce visibility, frequently obliterating detail altogether. For each 1% loss of contrast, 10% to 15% more light is needed to achieve equal visibility. Altering the position of the viewer or the task may eliminate the veiling reflections, or alternative directional lighting may be used to improve the contrast.

## EQUIVALENT SPHERE ILLUMINATION (ESI)

EQUIVALENT SPHERE ILLUMINATION (ESI) evaluates the way in which lighting systems affect task visibility, rather than just expressing the amount of light ("raw" footcandles). Sphere illumination refers to both the quantity and quality of light on a task. In the center of an evenly illuminated sphere in which light falls on a task equally from all directions, there is minimal loss in visibility from veiling reflections. From 45 to 250 "raw" footcandles, depending on the source location and direction of light, may be needed in an actual installation to produce the same visibility as 50 ESI footcandles provides. The better the quality of light, the fewer "raw" footcandles are needed for equal ease in seeing.

The ESI of a visual task in a real environment is the equivalent illuminance produced by a sphere which makes the task as visible in the sphere as it is in the real environment.

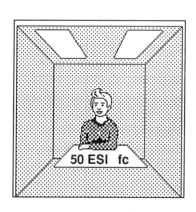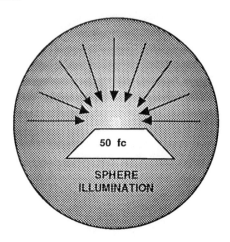

In an office, 50 ESI fc provides the same task visibility as 50 fc of sphere illumination.

Fig 1-12

ESI accounts for luminaire photometry, dimensions of a space, room reflectances and viewing position. Because of its complexity it is usually calculated on a computer. Lightolier publishes ratings for high performance troffers and task lights, as below.

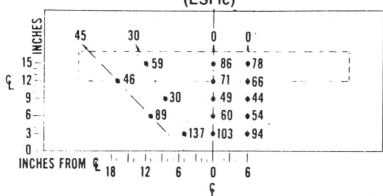

**Location and Viewing Direction for intital ESI-fc measurements on 60" x 24" work surface. Luminaire location shown in broken lines.**

Fig 1-13

## LIGHTING TECHNOLOGY

Any artist and designer must have both tools and materials to work with. The lighting designer's tools are his/her computer, pencil, calculator and mind. His/her primary materials, in addition to the light itself, are the fixtures, lamps and lenses that make up the luminaires themselves.

### COLOR AND LIGHT SOURCES

Color is a major factor in the emotional effect of any space. Without light, however, there is no color. Color is another and very important dimension of lighting design.

There are two aspects of color recognition: Light Source Color which is the spectral characteristics of the light falling on an object and Object Color which is the reflectance characteristics of an object. We see "color" because an object selectively reflects a certain portion of the light falling on it.

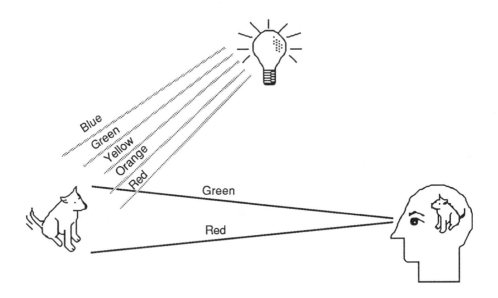

Fig. 1-14

Since white light consists of energy radiated throughout the visible spectrum, complementary colors (red/green, blue/orange) can be seen under it. If, however, a green leaf on a red apple were lighted with red wavelengths only, the leaf would appear without color or "black". If the apple were lighted with green light, it would appear "black" but the leaf would be perceived as green. Object colors, that is, paint, pigment or dye, function as selective reflectors. If a color is not in the light source, though, it cannot be seen in the object.

## SPECTRAL ENERGY DISTRIBUTION CURVES

The color composition of a light source can be drawn by plotting the amount of energy at each wavelength. This is a Spectral Energy Distribution Curve.

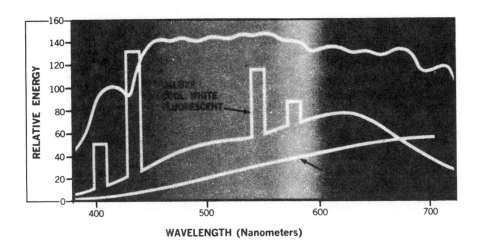

Fig 1-15

The above illustration shows relative spectral distributions for an incandescent source, a Deluxe Cool White Fluorescent lamp and noon sunlight with 400 nanometers (wavelength) being the blue end and 700 nanometers the red end of the spectrum.

The SED curve will indicate the color rendering properties of a light source. Incandescent sources are high in red and orange radiation, hence enhance "warm" colors such as red and orange and make dull or grey, "cool" colors such as blue or green. Because of their phosphor coatings, fluorescent and high intensity discharge lamps come in a variety of "whites" allowing for a choice of warm or cool color enhancement. See the section on light sources for more information.

## COLOR TEMPERATURE

A piece of metal heated to a high enough temperature will give off light as when a needle is sterilized by holding it over a flame. The CIE Chromaticity Chart below indicates the characteristics of a piece of metal in physics known as a theoretical "black body," heated to various temperatures measured in degrees Kelvin. The Kelvin (K) value indicates the degree of whiteness. The higher the temperature, the whiter the light.

Incandescent sources fall between approximately 2750K and 3200K, at the "warm" end of the spectrum. Because the fluorescent and high intensity discharge lamps do not generate light by means of incandescing, they do not fall on the black body line. Therefore, they have "correlated" color temperatures indicating the nearest point on the black body line,

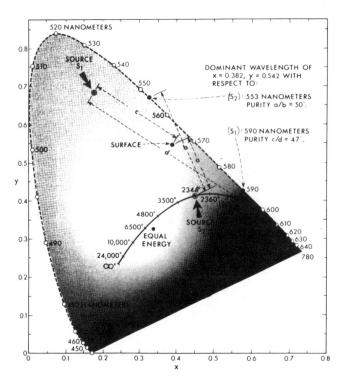

**Black body locus on C.I.E. chromaticity diagram.
Fig 1-16**

## COLOR RENDERING INDEX

Light sources may appear the same when lighted but unless their spectral composition is the same, objects colors will appear different. The Color Rendering Index is an indication of how similarly the color of an object is rendered by a non-incandescent source relative to a specific Kelvin temperature on the black body line. The higher the CRI, the better ("more natural") colors will appear. The maximum CRI is 100.

For instance, Warm White fluorescent, like tungsten halogen (an incandescent source), relates to a color temperature of 3200K, but because of its spectral make-up, there is a substantial difference between the perceived object color under the fluorescent source and under the "black body" source, resulting in a CRI or only 55.

Deluxe Cool White fluorescent has a CRI of 85, which is better, but it relates to a cooler, bluer part of the spectrum around 4100K. If a cool atmosphere is desired, then the higher the CRI relative to 4000K, the better colors will appear, but if a warm atmosphere is desired, sources relative to 3600K or below are more suitable.

**PERCEIVED COLOR EFFECTS FROM LAMPS**

| Lamp | Correlated Color Temperature | CRI | Lighted Appearance | Object Colors Enhanced | Object Colors Dulled |
|---|---|---|---|---|---|
| **Incandescent** | 2750° to 3200° | 89-99 | Yellowish white | Warm colors | Cool colors |
| **Fluorescent** | | | | | |
| Cool White | 4400° | 67 | White | Blue, yellow, orange | Red |
| Warm White | 3100° | 55 | Yellowish white | Yellow, orange | Red, blue |
| Cool White Deluxe | 4100° | 85 | White | All nearly equal | None |
| Warm White Deluxe | 3000° | 77 | Yellowish white | Red, yellow, orange | Blue |
| **HID** | | | | | |
| Clear Mercury | 5700° | 22 | Blue-green | Yellow, green, purple | Red, Orange |
| Deluxe Mercury | 3900° | 47 | Purplish white | Orange, yellow, purple | Deep reds |
| Deluxe Warm White | 3300° to 3600° | 47 | Yellowish white | Orange, yellow, purple | Deep reds |
| Metal Halide | 4700° | 53 | White | Orange, yellow, blue | Deep reds |
| High Pressure Sodium | 2100° | 22 | Yellow-orange | Yellow, orange | Green, Blue |

Fig 1-17

## LIGHT CONTROL

Light travels in a straight line until it strikes a surface. It is then modified by REFLECTION, TRANSMISSION, REFRACTION or ABSORPTION. Other possible modifications such as POLARIZATION, DIFFRACTION or INTERFERENCE which may occur are of minor concern in luminaire design.

## TYPES OF LIGHT MODIFICATION

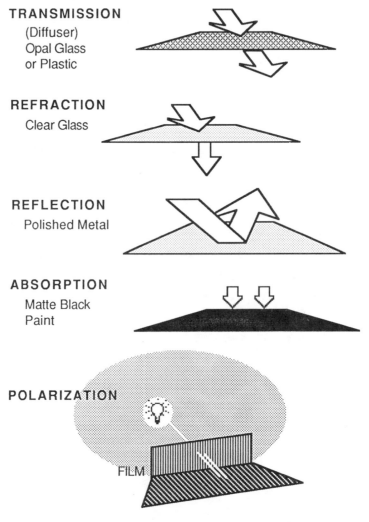

Types of Light Modification. Fig 1-18

A "bare lamp" luminaire is virtually 100% EFFICIENT, but it is not usually EFFECTIVE. An effective luminaire is one that directs the light output from the lamp into the zone where it is wanted and keeps it out of the zone where it is not wanted.

Light may be controlled or redirected as the luminaire designer wishes by using one or a combination of the following principles.

## TRANSMISSION

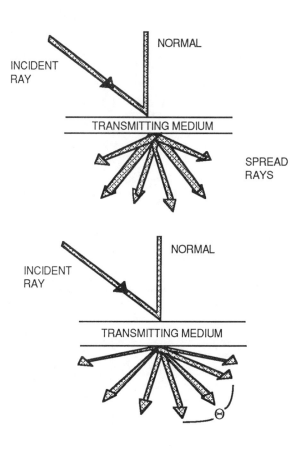

Fig 1-19

## DIRECT TRANSMISSION

DIRECT TRANSMISSION occurs with transparent materials which can be seen through, such as clear glass. They absorb the least amount of light in its passage through the medium.

## SPREAD TRANSMISSION

SPREAD TRANSMISSION occurs with translucent materials in which light passing through emerges at angles wider than the angle of incidence but the general direction of the beam remains the same. The light source is perceptible.

## DIFFUSE TRANSMISSION

DIFFUSE TRANSMISSION through such diffusing materials as opal glass or plastic, scatters light passing through in all direction and obscures the image of the light source. Diffusers usually transmit 40% to 60% of the incident light, but the optical system is usually greater than this due to inter-reflections.

# REFRACTION

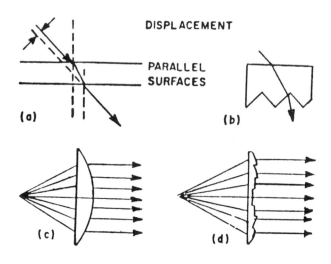

(a). Ray displaced. (b) Ray changed in direction. (c) Plano Convex Lens (d) Fresnel Lens.

Fig 1-20

REFRACTION occurs when a beam of light is "bent," that is, when the direction is altered in its passage from one medium to another. A ray of light is bent when it leaves air and enters a glass or plastic lens because of the difference in optical density of air and the lens material.

PRISMS are transparent forms with straight non-parallel sides that function as refractors. They are used to lower luminaire brightness or redirect light into useful zones.

A LENS is a refractor with one or more curved surfaces. Different kinds of lenses can cause parallel rays to converge or diverge, focusing or spreading the light. Complex optical systems may consist of several lenses used together.

The FRESNEL LENS is a type of convex lens that has its curved surface set back in a series of steps to reduce the overall thickness. It is generally used to bend light from a source for better utilization. Low brightness can be achieved by making the risers "dark" (which does not affect the beam control).

## REFLECTION

### *SPECULAR REFLECTION*

SPECULAR REFLECTION results from a shiny, highly polished or mirror surface. A beam of light is reflected at an angle equal to that at which it arrives, that is, the angle of reflection equals the angle of incidence. The smaller a source -- the more it approaches a theoretical "point" source -- the more precise the control of the reflected beam will be.

### *SPREAD REFLECTION*

SPREAD REFLECTION breaks up a beam of light in the general direction of the angle of reflection, spreading it somewhat, because of minute variations on the reflector surface, Specular reflectors are sometimes "roughed up" to provide a slight degree of diffusion for a softer, luminous effect or to hide filament striation or lamp irregularities.

### *DIFFUSE REFLECTION*

DIFFUSE REFLECTION is characterized by light leaving a surface in all directions as with white plaster or flat white paint.

## CONTROL OF THE BEAM

The direction of light from a luminaire is often controlled by reflectors. The shape of a reflector contour determines the beam pattern.
A variety of distribution patterns are shown below.

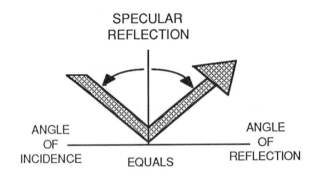

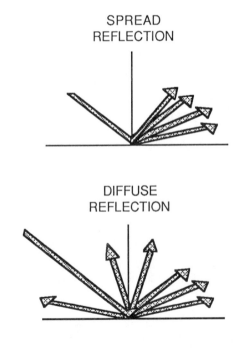

Reflectors may produce narrow to wide beams.
Fig 1-21

## ABSORPTION

There is always a subtractive interaction when light hits a surface, that is, some light is lost by ABSORPTION. On the other hand, absorption, as with baffles or louvers, is never complete either. Matte black paint offers nearly complete absorption and when combined with the shadows cast by parallel baffles as in a Multi-groove downlight, can create a very low surface brightness.

## POLARIZATION

POLARIZATION refers to light waves vibrating in one plane only. Polarizing materials can be used to improve task visibility by reducing reflected glare.

## LUMINAIRES

The luminaire is the aggregate of lenses, lamps and reflectors assembled and placed in the building or physical location dictated by the design. We will, of course, dissect the luminaire to better understand how it works. Any designer must know both the inner workings and the effects of the lighting instruments utilized in a design. It is normal to select a luminaire by the desired effect and it is not unusual to specify a modification of a selected luminaire to make a desired effect or even to design a completely new luminaire to achieve a design end.

### CLASSIFICATION OF LUMINAIRES

A direct system (downlighting) is the most efficient in delivering the maximum amount of light on a work surface because it is less dependent on reflecting surfaces than any other system. However, a direct system may also produce the greatest amount of distracting brightness contrast with the ceiling and the greatest glare and deepest shadows. The larger the upward component of light, the more critical the ceiling reflectivity and the lower the coefficient of utilization (CU) or efficiency of the luminaire.

The distribution characteristics of luminaires are one of the most fundamental aspects of lighting elements in a design. The selection of the control of artificial lighting is the primary way one has of predicting the effect of an installation. Both the aesthetics and the technology of light are integrated in this concept. The selection and placement of fixtures, the selection of lamps, the design and testing of luminaires--all focus on the manner and efficiency of distribution. Based on their lighting distribution, the IES classifies luminaires as follows:

## DIRECT

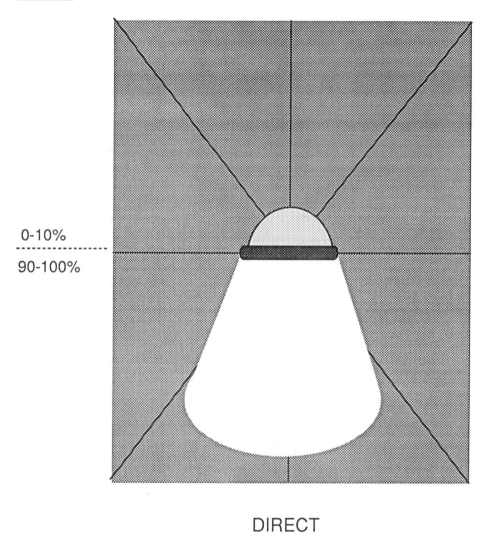

**Fig 1-22**

The direct pattern gives the designer the maximum control when lighting a task but it does not accommodate the ambient element.

## SEMI DIRECT

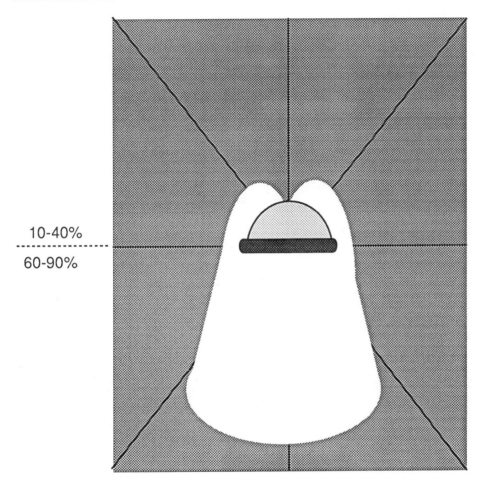

10-40%
- - - - - - - - - - - - - -
60-90%

**SEMI DIRECT**

**Fig 1-23**

This pattern includes the strengths of direct while providing an ambient or indirect element.

## GENERAL DIFFUSE

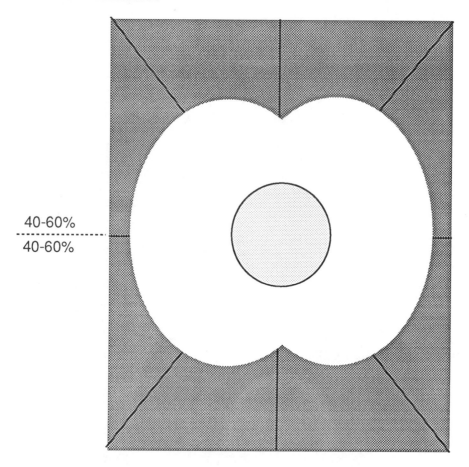

**GENERAL DIFFUSE**

**Fig 1-24**

This pattern achieves maximum ambient effect but maximizes glare and eschews task focus.

## DIRECT-INDIRECT

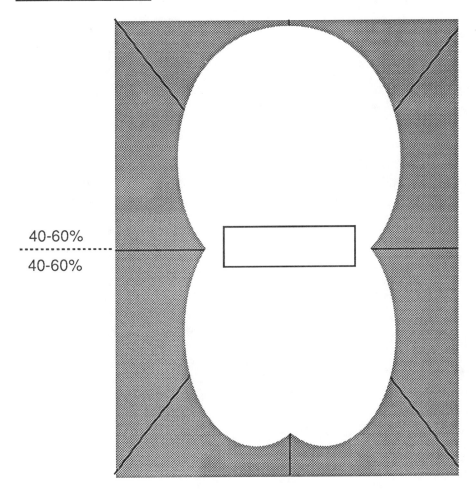

**Fig 1-25**

This pattern allows efficient energy use while maximizing task illumination.

## SEMI-INDIRECT

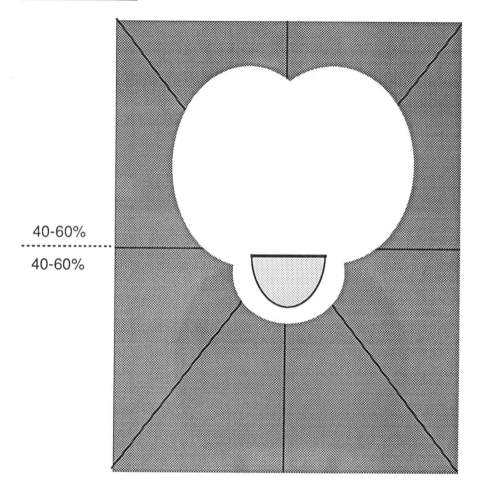

SEMI-INDIRECT

Fig 1-26

This pattern is good for aesthetic ambient effects while providing useable non-critical task illumination.

## INDIRECT

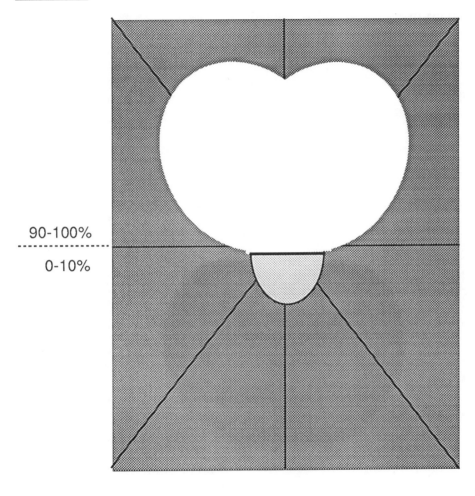

### INDIRECT

Fig 1-27

This is one of the most pleasing ambient patterns but is the most energy inefficient luminaire.

## LUMINAIRE SPECIFICATIONS

A luminaire is a complete lighting unit. It includes the housing, lampholders or sockets, control devices and lamps. A luminaire SPECIFICATION SHEET is the basic source of information on a luminaire. Along with a picture or drawing of the unit, dimensional data, description of materials, electrical requirements and ordering information, specification sheets should include some or all of such pertinent performance information as follows.

### PHOTOMETRIC DATA

Light from a luminaire can be measured by a photometer, a device that delineates the candlepower distribution of a luminaire. The candlepower distribution curve indicates the direction and magnitude of light intensity (luminous flux) from which footcandles may be calculated. Non-affiliated testing laboratories such as Independent Testing Laboratory (ITL), Electrical Testing Laboratory (ETL), Lighting Sciences Inc. (LSI) and Environmental Research Laboratories (ERL) provide photometric reports for the lighting industry.

Learning to read and use photometric data is an important task for the student of lighting design. If you don't know what effect a luminaire will have in a given environment the chance of achieving a predictable lighting design becomes remote. Without the luxury of creating a physical test model of a lighting installation before construction, there is no other source of design information than the manufacturer's photometric data sheets and the lighting design calculations that depend on them. Pay strict attention to mastering this area.

### POLAR GRAPH

The POLAR GRAPH is the commonly used diagram of candlepower distribution, in which all straight lines relate to a single point as in the ITL Report, below. Degrees from vertical are represented by lines radiating from a single point. Candelas are read on the vertical scale. With symmetrical distribution, only half of the pattern need be shown. Luminaires with an upward component require a 180° graph. Recessed luminaires supply no light above the ceiling, so they may be represented on a quadrant graph of 0° to 90°.

To determine the candlepower of the source plotted in the chart below, at say 30° from vertical, follow the straight radiating line marked 30° until it meets the curved line. Then read (or interpolate) the amount of candlepower directly to the left on the vertical line. Fig. 1-28 shows that a 100A19 lamp in this luminaire produces 950 candelas at 30° from vertical.  Candlepower at specific angles is also shown in tabular form to the left of the graph.

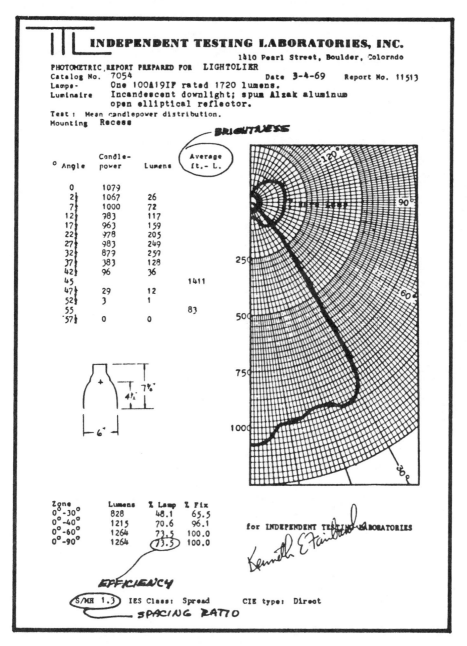

ITL Report on incandescent downlight. Fig 1-28

## EFFICIENCY

Efficiency is the ratio of lumens leaving a luminaire to the amount of initial lamp lumens. It is shown on the photometric report in the column marked % LAMP. For example, the downlight shown in Figure 1-28 has an efficiency of 73.5% ($\frac{1264L}{1720L}$ x 100 ) as shown in the 0° to 90° zone line.

## SPACING RATIO

A laboratory report will show the maximum ratio of SPACING TO MOUNTING HEIGHT (S/MH) on the report as indicated. This SPACING RATIO is the maximum ratio of the distance between luminaires to the height above the work plane that will provide relatively even lighting, that is, not more than 1/6 above or below the average level of illumination. The maximum spacing for a particular luminaire is determined by multiplying the mounting height by SPACING RATIO (SR).

Work surfaces are considered to be 30" above the floor in offices, schools, libraries, etc. When calculating spacing of luminaires in such spaces, subtract 30" from the ceiling height to find the mounting height. For example, in a room with a 9.5' high ceiling, luminaires with a SR 9 or S/MH of 1.2 will provide acceptably uniform illumination if they are spaced not more than 8.4' apart. (9.5' - 2.5' = 7'MH x 1.2SR = 8.4').

## NON-SYMMETRICAL DISTRIBUTION

Most fluorescent luminaires do not have symmetrical distribution, hence they are shown with readings taken in at least three vertical planes-- parallel to the lamps, perpendicular to them and at 45°. The parallel and perpendicular values are used to plot the curves shown on the polar graph. See the ITL report below. The candlepower readings are listed for 0° to 90° (it is a recessed unit).

## BRIGHTNESS

Most incandescent and HID downlights produce low apparent brightness at normal viewing angles, that is, 52° from vertical to horizontal. Fluorescent luminaires, however, frequently produce light at higher angles, hence brightness becomes an important consideration in luminaire choice. Average and maximum brightness in footlamberts (FL) for various angles is given on the photometric report as in Fig. 1-29. These brightness readings are the basis of calculating the VCP of a luminaire.

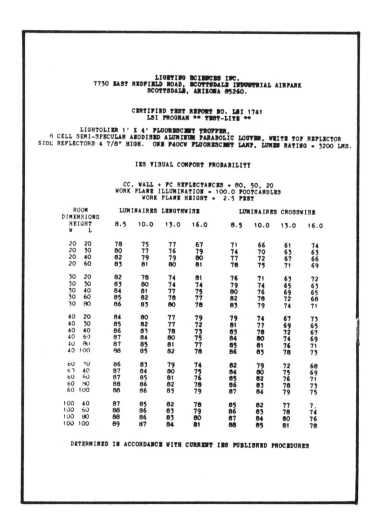

**ITL Report on fluorescent troffer. Fig 1-29**

## COEFFICIENTS OF UTILIZATION

The COEFFICIENT OF UTILIZATION (CU) is the ratio of lumens arriving at a work surface to total lamp lumens in the lighting system. Independent testing laboratories usually provide CU tables for luminaire manufacturers, thus guaranteeing objectivity. Factors that go into the CU are the efficiency of the luminaire, the size and shape of the room, the reflectance of the ceiling, walls and floor, and the candlepower distribution pattern. CU's provide a reasonably accurate and speedy method for calculating raw footcandle levels in a space by the Zonal Cavity Method.

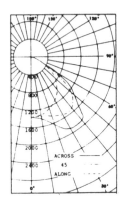

| | Ceiling Cavity Reflectance | | | | | | | | | | | | | | |
|---|---|---|---|---|---|---|---|---|---|---|---|---|---|---|---|
| | 80 | | | 70 | | | 50 | | | 30 | | | 10 | | 0 |
| | Wall Reflectance | | | | | | | | | | | | | | |
| Room Cavity Ratio | 50 | 30 | 10 | 50 | 30 | 10 | 50 | 30 | 10 | 50 | 30 | 10 | 50 | 30 | 10 | 0 |
| | Coefficients of Utilization | | | | | | | | | | | | | | |
| 1 | .72 | .70 | .68 | .71 | .69 | .67 | .68 | .66 | .65 | .65 | .64 | .63 | .63 | .62 | .61 | .60 |
| 2 | .65 | .61 | .58 | .64 | .60 | .57 | .61 | .59 | .56 | .59 | .57 | .55 | .57 | .55 | .54 | .52 |
| 3 | .58 | .54 | .50 | .57 | .53 | .50 | .55 | .52 | .49 | .53 | .51 | .48 | .52 | .49 | .47 | .46 |
| 4 | .52 | .47 | .43 | .51 | .47 | .43 | .50 | .45 | .42 | .48 | .44 | .42 | .47 | .44 | .41 | .40 |
| 5 | .46 | .41 | .37 | .46 | .40 | .37 | .44 | .40 | .36 | .43 | .39 | .36 | .42 | .38 | .35 | .34 |
| 6 | .41 | .36 | .32 | .41 | .35 | .32 | .39 | .35 | .31 | .38 | .34 | .31 | .37 | .34 | .31 | .29 |
| 7 | .37 | .31 | .28 | .36 | .31 | .27 | .35 | .30 | .27 | .34 | .30 | .27 | .34 | .30 | .27 | .25 |
| 8 | .33 | .27 | .23 | .32 | .27 | .23 | .31 | .27 | .23 | .31 | .26 | .23 | .30 | .26 | .23 | .22 |
| 9 | .29 | .24 | .20 | .29 | .24 | .20 | .28 | .23 | .20 | .27 | .23 | .20 | .27 | .22 | .19 | .18 |
| 10 | .26 | .21 | .17 | .26 | .21 | .17 | .25 | .20 | .17 | .25 | .20 | .17 | .24 | .20 | .17 | .16 |

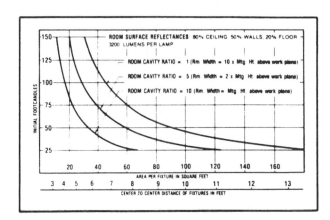

**Typical data allows calculations for either single or multiple luminaires.**

**Fig 1-30**

## HOW TO USE PHOTOMETRIC DATA

### LEVELS OF ILLUMINATION

Chapter Four explains how to calculate footcandle levels at a point or for an area from photometric data. Many specification sheets provide pre-calculated information or Quick Calculator charts such as those shown in Fig. 1-30 above. From left to right, the data given for this incandescent downlight are: 1. the candlepower distribution curve from the photometric report; 2. footcandle levels and beam spread for SINGLE UNIT (Point Method); 3. table of coefficients of utilization and 4. footcandle levels and spacing ratio for area lighting from MULTIPLE UNITS (Zonal Cavity Method). Conversion factors are also shown for use with different lamps and finishes.

### BRIGHTNESS

Glare is brightness which creates annoyance and discomfort. High brightness in the 60° to 90° zone is the cause of direct glare. Light rays in the 0° to 30° zone are generally the source of veiling reflections. The zone between 30° and 60° produces the most effective, that is, most useful light, the light that most helps us to see the task better.

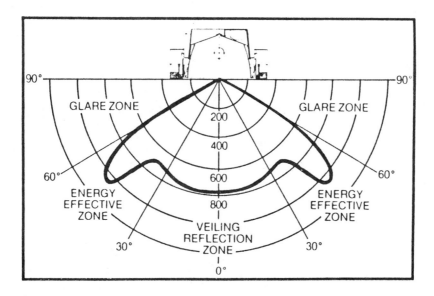

The 30°-60° zone is the energy effective zone. Fig 1-31

According to the Illuminating Engineering Society, direct glare will not be a problem IF ALL THREE of the following conditions in a installation are met:

1.   The VCP is 70 or more

2.   The ratio of maximum-to-average footlamberts does not exceed 5 to 1 (preferably 3 to 1) at 45, 55, 65, 75 and 5 degrees from nadir (lowest point directly under the fixture) crosswise or lengthwise.

3.   Maximum luminaire brightness does not exceed the following values:

| Angle above Nadir | Maximum Footlamberts |
|---|---|
| 45 | 2250 |
| 55 | 1605 |
| 65 | 1125 |
| 75 | 750 |
| 5 | 495 |

Performance specifications such as in the chart below state the brightness of a luminaire in footlamberts, the VCP and the relative amount of flux in each of the three zones as an aid in luminaire selection.

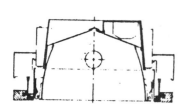 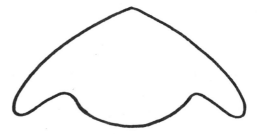

## PERFORMANCE SPECIFICATION

|  | 1' x 4' 1-40W |
|---|---|
| **COEFFICIENT OF UTILIZATION** Reflectances: 80%, 50%, 20% Room Cavity Ratio = 1  Not Less Than | .75 |
| **EFFICIENCY** — Not Less Than | 70% |
| **MAXIMUM AVERAGE BRIGHTNESS** Lengthwise or Crosswise at 65°  Not More Than | 290 fl |
| VCP (at rear of room) Lengthwise/Crosswise (60' x 60' x 10' Room)  Not Less Than | 85/82 |
| **LUMINAIRE LIGHT FLUX** 0°-30° Zone — Less Than 30°-60° Zone — Greater Than 60°-90° Zone — Less Than | 28% 62% 11% |

**Performance specification for the 1 Lt. 40W HVP 3. Fig 1-32**

## GLARE

### CONTROL OF LUMINAIRE BRIGHTNESS

Direct viewing of either the lamp or the interior of a luminaire can be a source of discomfort glare. Devices for reducing luminaire brightness to a comfort range in normal viewing angles (40 degrees and above) include parabolic cone reflectors, lenses, baffles and louvers.

These elements and their availability in a selected luminaire are critical in creating both an aesthetic and physically comfortable lighting design. While the placement of a luminaire is important in the elimination of glare, if a luminaire has no inherent glare control the effectiveness of the placement and the efficiency of the installation can be greatly impacted.

A SPECULAR CONE REFLECTOR (shiny black, gold or natural Alzak) is a section of a parabolic reflector that redirects the light incident upon it straight down, thereby eliminating uncomfortable brightness at high angles (above 45 degrees).

A LENS made of either plastic or glass, intercepts as much light as possible and redirects it into the more useful zones. Luminaire brightness is reduced because all light rays pass through the lens, blocking direct viewing of the lamps.

A BAFFLE is an opaque element of wood, metal or plastic that prevents direct viewing of a lamp or reduces the brightness of a lens or diffuser. Horizontal baffles mounted one above the other reduce brightness even more than vertical baffles because each one throws a shadow on the one below as in a multi-groove baffle.

A LOUVER is an assembly of vertical baffles arranged parallel, in a grid or in a concentric pattern to block the view of high luminaire brightness above 45 degrees. SPECULAR PARABOLIC LOUVERS combine reflective principles with a grid baffle to redirect light downward, producing a very low brightness appearance.

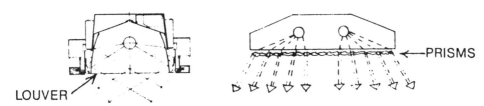

**FLUORESCENT**

LOUVER

PRISMS

**INCANDESCENT**

SPECULAR

BAFFLE

**Light Control Devices applied to luminaires. Fig 1-33**

## SELECTING A LUMINAIRE

After illumination levels are determined, there are trade-offs to be weighed when selecting luminaires. They include: economic considerations, suitability of installation, efficiency and lighting effectiveness. The lighting effectiveness of a luminaire may result in the need for fewer luminaires, hence lower installation and operating costs, than a more efficient luminaire that does not control brightness as well. Higher efficiency must be weighed against comfort, distribution and glare. The information given on a luminaire specification sheet should contain all the necessary lighting performance data and details other than price to make an informed decision.

## SUMMARY

The design medium, as we have discovered, includes one of the most complex assemblies of design elements of any art form. The elements range from paint pigments to electronic control devices. The more comprehensive the knowlege of the designer with regard to these design elements, the better the design result will be.

I must also re-emphasize the systematic nature of the lighting design process. Remember that both the intention of the designer and the perception of the observer converge in the realized design. Lighting is never the subjective vision of the designer alone. It depends on a complex interplay of technical and psychological elements.

# Chapter 2

## DESIGN PRINCIPLES

### THE ART AND SCIENCE OF LIGHTING

Lighting design is an art and a science. As a science, the amounts of illumination needed and certain aspects of the quality of light have been quantified. As an art however, to attach numbers is meaningless because light is an experience of the SENSES. It is not an INTELLECTUAL experience. Lighting in a space is a positive force that can motivate people to be active, relaxed, productive, lively or depressed. Lighting should make people important. It should create an atmosphere pleasing to the occupants whether in an office, store, showroom or home. Lighting should provide visibility, character and mood as well as relate harmoniously to the space in which it is used.

Lighting design is the process of integrating, in a unique way, the art and science of human perception with the art and science of human technology. The result is a very complex system that varies in time in a way that can be extremely exciting. The complex and temporal nature of lighting is one of the least understood of its many variables. The time of day and year, the age and mood of the observer, the use and location of the architectural space being experienced--all these elements come into play with the intention of the lighting designer and result in a myriad of totally unique experiences of the environment. Because of this complexity, lighting design can be one of the most creative areas of all of architecture. In this chapter we will introduce the elements of design with which we work.

## PRINCIPLES

### EFFECT ON ARCHITECTURE

Light is as much a "building material" as steel or concrete. Although such structural components are needed to enclose a space, it has no real existence for an individual until it is seen and it registers in his consciousness. Light defines space; reveals texture; shows form; indicates scale; separates functions. Good lighting makes a building look and work the way the architect intended at all hours of day and night. It contributes to the character, to the desired attitude toward form and space, and to the effective functioning of that space. Lighting is dynamic. Change the lighting and the world around us changes.

### EFFECT ON INTERIOR DESIGN

Light is invisible until it strikes a surface and molds our environment. How light strikes, the angle, the quality, the intensity, all combine to render objects in different ways.

**Fig 2-1**

The vital relationship between light and color can enhance or destroy the most carefully worked out color scheme. Knowing the effect of light on a surface, the designer can choose the most suitable appearance to accomplish his purpose. Light models, creates coherence, gives unity. Light can make or break a space both functionally and aesthetically.

## PLANES OF BRIGHTNESS

Where brightness occurs in a space establishes the character or mood of that space.

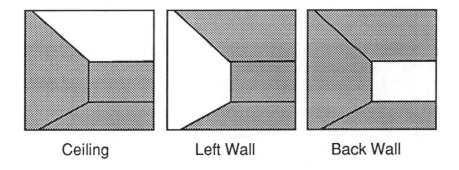

Ceiling       Left Wall       Back Wall

**The eye is drawn to the brightest surface in the room. Fig 2-2**

A ceiling left in shadow creates a secure, intimate, relaxing environment suitable for lunges, leisurely dining and board rooms. High brightness on the ceiling creates a bright and cheerful or efficient and work-like atmosphere good in coffee shops, classrooms and kitchens. With the ceiling in shadow, brightness on the vertical planes of a space draws attention to the walls, expanding space visually; good for galleries, merchandising, lobbies. A pattern of varying levels of brightness can indicate direction and lead people through a space.

## GLITTER AND SPARKLE

Pinpoints of brightness from small exposed filaments or multiple reflection from crystal, chrome or other shiny surfaces create a scintillating effect that heightens awareness. Glitter and sparkle add gaiety, sophistication, elegance and festivity to a space. Whether used for dining, dancing or merchandising, glitter can become glare without sufficient background lighting to soften the contrast. There is only a fine line between stimulating points of brightness and discomfort glare.

**Fig 2-3**

## LIGHT AND SHADOW

An evenly illuminated space is similar to an overcast day--dull, monotonous and boring. Variations in brightness and the interplay of light and shadow add variety to a space, provide visual relief and a sense of excitement. Scallops of light on a wall from nearby downlights, shadows on the ceiling of varying sizes and shapes from an uplight under a plant or a narrow beam of light highlighting a small sculpture, create areas of visual interest that give character and individuality to a space. Highlight provided by accent light creates focal points that can direct attention or communicate an idea. A lighting scheme should not be so exciting as to overwhelm or destroy the interior design, nor so placid as to make a room dull and uninteresting.

**Fig 2-4**

## MODELING

Shadows are essential for perceiving dimensionality. Three dimensional objects lighted from directly in front appear flat, but when lighted from the side, assume depth and roundness. The deep shadows created by strong source from one side only cause more distortion but because of the high contrast add to its dramatic impact. Free standing objects, such as sculpture, lighted from two directions with different intensities or tints of color will appear fully three dimensional.

Fig 2-5

## LIGHTING DESIGN CONSIDERATIONS

The most important factor in a space is the people who will occupy it, use it, live in it. People are not automatons; therefore, the psychological and emotional effects of an environment are of equal importance with the physiological.. Not only should good quality light be provided to "see by" but also to "feel by."

The process the designer uses and the elements that the designer considers while making and realizing a design should be based on the entire environmental and psycho-physical system that interact with the observer to create a lighting installation. The design process itself should reflect the attitudes and elements of design that are essential. A good designer *always* touches base with the *entire* scope of the environmental design and *never* leaves it to chance.

The factors that should be considered in lighting any space are:

THE LIGHTING DESIGN PROCESS

1. The SITUATION; is it a working. viewing, circulation or living space?

2. The FUNCTION; what will people do in the space? Type, study, eat , sew, or buy and sell?

3. The QUANTITY AND QUALITY OF LIGHT needed to perform tasks.

4. The ARCHITECTURE and DECOR.

5. The ATMOSPHERE or psychology of the space.

6. The RELATIONSHIP to adjacent areas.

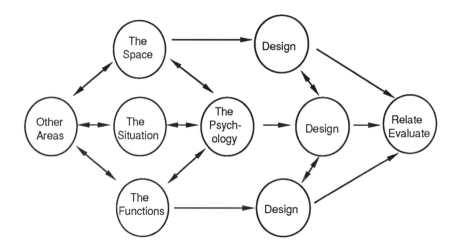

**Spatial, functional and psychological criteria are interrelated.**
**Fig 2-6**

The diagram above shows a method of integrating spatial, functional and psychological criteria to arrive at a suitable design conclusion. The final evaluation should include economic considerations such as behavioral benefits, added value, safety, and available capital as well as energy criteria and integration with thermal and acoustical needs.

The process of evaluating all of these design elements to produce a design document from which the contractor or builder can realize the actual lighting installation is complex but similar to the way a design professional proceeds in any other discipline. In this book are included a series of forms, templates, examples and formulas for the evolution, analysis and communication of the design.

You should select those elements that are most helpful to you and expand and adapt them to your own situation. At the same time you must remember that there is a common language and symbology that simplifies and clarifies your ideas and their communication to others. Take heed not to deviate from the standard too far or you risk being misunderstood in important details.

## KINDS OF LIGHTING

TASK LIGHTING is illumination from nearby sources that allows the performance of visual tasks. It is localized light to work by.

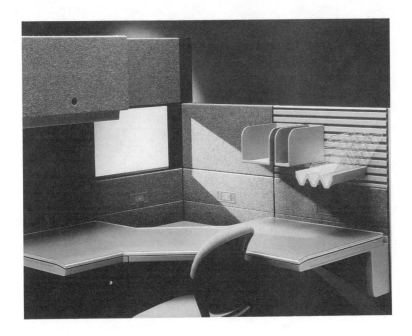

Fig 2-7

Task lighting is one of the most important elements of commercial lighting design. Its point is to provide the necessary illumination for the worker to perform his/her task. There are many approaches to this job from specialized lighting fixtures to integrated lighting systems.

ACCENT LIGHTING is directional light intended to emphasize a particular object or draw attention to a part of the field of view.

Fig 2-8

Accent lighting is an important lighting element for the architect for it allows the appropriate emphasis of a design element or detail which can enhance or underscore a theme or essence of his/her design. It is also a way to focus attention on signage or art in an environment.

Chapter Two   Design Medium   53

GENERAL or AMBIENT LIGHTING is background or fill light in a space that reduces harsh contrast between pools of localized tasks or accent or that supplies a substantially even level of illumination throughout an area.

**Fig 2-9**

Just as the background paper in a this book plays an important role in our ability to read the words printed on it, the ambient light of an architectural space provides a ground upon which the architect and lighting designer can practice his/her art. This is an often underdeveloped element of designs and must be given its appropriate due.

# TECHNIQUES

## OVERVIEW

Lighting design techniques are essentially design elements or solutions that can be utilized in many design contexts. They are analogous to standard letters that can be edited and modified by the user for specific situations, thereby saving time and giving the writer a head start. I have included a series of standard design techniques in this section for your use. Remember, these elements have to be customized for your specific application. The aim is to utilize techniques to get the design concept realized. Don't substitute technique for substance in your design. If you don't start with a well worked out design concept all the technical knowledge in the world will not help you.

## SURFACES

The vertical architectural element or surface is the most common element that requires lighting. There are many things that can be done to enhance surfaces both vertical and horizontal. Some of those techniques will be introduced in this section.

### LIGHTING THE VERTICAL SURFACE

The lighting of vertical surfaces, that is, walls, panels, draperies, dividers or wall displays, can be continuous as with wall lighting or discontinuous as with accent or display lighting. A smooth even distribution of light over a wall is called "WALL WASH". Wall lighting may be from incandescent, HID or fluorescent sources, the selection being based on the size of the space and the evenness of the brightness, the intensity and the textural emphasis desired. Accent or display lighting is usually incandescent because of its suitability to directional control.

TEXTURE RENDITION LIGHTING can emphasize texture and surface irregularities if it is installed close to a wall ("grazing" light) as with light sources concealed behind a baffle or in a trough next to the wall, or it can minimize texture-- and faulty seams or plaster work--if it is installed at some distance form the wall as with wall washer luminaires. The farther out from the wall the luminaires are, the "flatter" the wall will appear and the less its surface variations and texture will be noticed. Wall washers may be recessed, surface or track mounted.

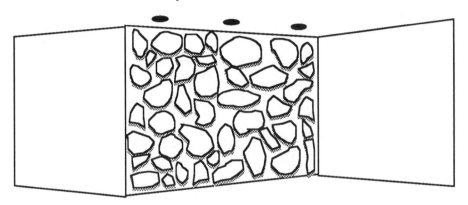

**Emphasizing Texture. Fig 2-10**

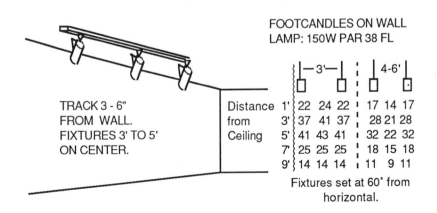

**Track Lighting. Fig 2-11**

More complete data is also available for other luminaires from various catalogs as shown below.

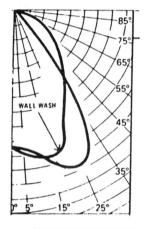

For even wall washing use maximum of 2 to 1 spacing (distance between fixtures ÷ distance to wall).

### FOOTCANDLES ON WALL FROM MULTIPLE UNITS
### CLEAR CONE, 150W A-21

**UNITS 3 FEET FROM WALL**

| DISTANCE FROM CEILING, FT. | ⊢— 3' —⊣ | | | ⊢— 4' —⊣ | | | ⊢— 5' —⊣ | | | ⊢— 6' —⊣ | |
|---|---|---|---|---|---|---|---|---|---|---|---|
| 1  | 21 | 20 | 21 | 18 | 16 | 18 | 16 | 10 | 16 | 14 | 8  | 14 |
| 2  | 39 | 38 | 39 | 32 | 28 | 32 | 28 | 18 | 28 | 26 | 14 | 26 |
| 3  | 36 | 37 | 36 | 27 | 28 | 27 | 23 | 23 | 23 | 20 | 17 | 20 |
| 4  | 33 | 33 | 33 | 26 | 24 | 26 | 22 | 19 | 22 | 19 | 16 | 19 |
| 5  | 29 | 30 | 29 | 23 | 24 | 23 | 19 | 19 | 19 | 16 | 15 | 16 |
| 6  | 26 | 26 | 26 | 19 | 21 | 19 | 16 | 17 | 16 | 13 | 14 | 13 |
| 7  | 23 | 24 | 23 | 18 | 18 | 18 | 14 | 15 | 14 | 11 | 13 | 11 |
| 8  | 21 | 23 | 21 | 17 | 18 | 17 | 14 | 14 | 14 | 11 | 11 | 11 |
| 9  | 20 | 22 | 20 | 17 | 17 | 17 | 14 | 13 | 14 | 11 | 11 | 11 |
| 10 | 20 | 20 | 20 | 15 | 16 | 15 | 13 | 12 | 13 | 11 | 10 | 11 |

Footcandle values are averaged and rounded off and are based on a minimum of five units.
Conversion Factors: 100W A-19 (Clear): F.C. x 0.6; 100W A-19 (Gold): F.C. x 0.55, 150W A 21 (Gold) F.C x 0.9.

**Wall washer data shown for 3' from wall. Note the effect distance and spacing have on footcandles.**
**Fig 2-12**

## CONTINUOUS FLUORESCENT STRIP SYSTEMS

CONTINUOUS FLUORESCENT STRIP SYSTEMS are particularly effective when used to make ceilings appear to "float." Because of the diffuse nature of the light with subsequent lack of directionality inherent in fluorescent lighting, the wall brightness drops off quickly with distance from the ceiling; hence most fluorescent systems such as the Trough Systems are not primarily wall washers and are generally best suited to 10' ceilings or lower. A typical illumination chart for Perimeter Trough Lighting showing installation variations follows:

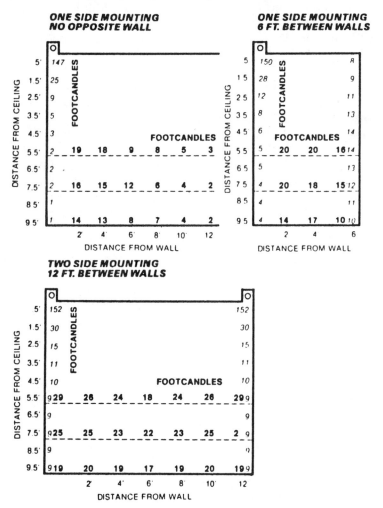

**Perimeter 2 Louver vertical and horizontal footcandle data.**
**Fig 2-13**

In runs that can be seen lengthwise, such as those that wrap around a corner where the viewer is at right angles to the trough, louver shielding should be used to prevent direct sight of the high interior luminance of the trough.

## ACCENT LIGHTING

An accented element should be at least three times as bright as its background for discernible contrast. For greater emphasis, as in lighting a feature display or store manikin, the ratio should be at least five to one. The darker the object being lighted, the more illumination it requires to stand out from its background. If footcandle charts are not available, use the Point Method of calculation to determine footcandle levels.

Discontinuous wall wash lighting can emphasize a section of a wall, good for highlighting large-scale wall hangings. Localized sources that produce a confined beam of light serve to focus attention on smaller pictures, signs, graphics, displays, etc. Aiming the luminaire beam 30 degrees from the vertical will prevent reflected glare -- especially critical with glass-covered or glossy elements -- and will not create distracting shadows from a frame. In most instances, the 30 degree aiming angle also eliminates the possibility of someone walking through the beam of light. More horizontal beams may produce reflected glare; more vertical beams, distorting shadows.

## FRAMING PROJECTORS

FRAMING PROJECTORS have a series of lenses and adjustable shutters that allow for shaping the beam. They are usually used to illuminate a picture precisely without spill light on the adjacent wall. They are also sometimes used to project a shaped beam pattern of light on a table or wall for drama. The chart below gives data for beam size and footcandle levels.

| MAXIMUM BEAM COVERAGE AND ILLUMINATION ON VERTICAL PLANE | | | | | | | | | | | | | | |
|---|---|---|---|---|---|---|---|---|---|---|---|---|---|---|
| AIMING ANGLES: 45°, 60° FROM HORIZONTAL. SHUTTERS FULLY OPEN. | | | | | | | | | | | | | | |
| DISTANCE (D) | 24" | | 30" | | 36" | | 42" | | 48" | | 54" | | 60" | |
| AIMING ANGLE (A) (from horizontal) | 45° | 60° | 45° | 60° | 45° | 60° | 45° | 60° | 45° | 60° | 45° | 60° | 45° | 60° |
| DISTANCE (C) (Min.) | 8" | 17" | 11" | 21" | 13" | 26" | 16" | 30" | 18" | 35" | 21" | 40" | 23" | 44" |
| WIDTH (W) (Max.) | 20" | 25" | 25" | 31" | 29" | 38" | 34" | 44" | 39" | 50" | 44" | 56" | 49" | 63" |
| LENGTH (L) (Max.)* | 37" | 65" | 46" | 82" | 56" | 98" | 65" | 114" | 74" | 131" | 84" | 147" | 93" | 163" |
| F.C. ON BEAM AXIS | 63 | 22 | 41 | 14 | 28 | 10 | 21 | 7 | 16 | 6 | 13 | 5 | 10 | 4 |

*Length (L) on the point where illumination drops to 10% of illumination on beam axis.

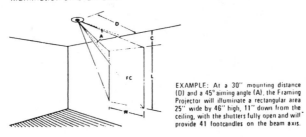

EXAMPLE: At a 30" mounting distance (D) and a 45° aiming angle (A), the Framing Projector will illuminate a rectangular area 25" wide by 46" high, 11" down from the ceiling, with the shutters fully open and will provide 41 footcandles on the beam axis.

| MAXIMUM BEAM COVERAGE AND ILLUMINATION ON HORIZONTAL PLANE | | | | | | | | |
|---|---|---|---|---|---|---|---|---|
| AIMING ANGLE: 0° FROM VERTICAL SHUTTERS FULLY OPEN. | | | | | | | | |
| DISTANCE (H) | 4' | 5' | 6' | 7' | 8' | 9' | 10' | 11' | 12' |
| SIDES (S) (Max.) | 41" | 51" | 61" | 71" | 81" | 91" | 101" | 111" | 121" |
| F.C. ON BEAM AXIS | 41 | 27 | 19 | 14 | 11 | 9 | 7 | 6 | 5 |

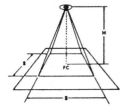

EXAMPLE: At a 6' mounting height (H) and an aiming angle of 0° from vertical, the Framing Projector will illuminate a square area 61" by 61" with the shutters fully open and will provide 19 footcandles on the beam axis.

**Maximum beam coverage and illumination of a vertical and horizontal plane.**

**Fig 2-15**

## DESIGNING THE WORKSPACE

Aside from the aesthetics of lighting the architectural space the primary job of the designer is the effective illumination of the workspace. Most workspaces today are poorly lighted. Often the primary problem is glare either direct or indirect on CRT screens. It is important to carefully consider all the work circumstances that will be encountered in the environment you are developing and to create a humane lighting design that enhances rather than inhibits effective, efficient and joyful work.

## TASK VISIBILITY

A visual task is the seeing job that must be performed in combination with office work, food preparation, handcraft or any activity that requires close visual attention and discrimination among details. As described earlier, the size and contrast of the details of a task, its brightness (luminance) and the time required to perceive the visual message are the factors that influence task visibility. Up to a certain point, these factors improve as the quantity of illumination increases. The IES publishes recommended footcandle levels for a variety of visual tasks. The chart under LIGHTING THE TASK (page 65) indicates IES recommended values for footcandles for selected office tasks based upon the procedure in use since 1958.

In 1979 the IES established a new procedure for recommending footcandle levels. The new method uses a footcandle range approach accompanied by weighting factors as shown in Figure 2-16.

| a. General Lighting Throughout Room ||||||
|---|---|---|---|---|---|
| Weighting Factors || Illuminance Categories |||
| Average of Occupants Ages | Average Room Surface Reflectance (per cent) | A | B | C |
| Under 40 | Over 70 | 20 | 50 | 100 |
|  | 30–70 | 20 | 50 | 100 |
|  | Under 30 | 20 | 50 | 100 |
| 40–55 | Over 70 | 20 | 50 | 100 |
|  | 30–70 | 30 | 75 | 150 |
|  | Under 30 | 50 | 100 | 200 |
| Over 55 | Over 70 | 30 | 75 | 150 |
|  | 30–70 | 50 | 100 | 200 |
|  | Under 30 | 50 | 100 | 200 |

| b. Illuminance on Task ||||||||| 
|---|---|---|---|---|---|---|---|---|
| Weighting Factors ||| Illuminance Categories ||||||
| Average of Workers Ages | Demand for Speed and/or Accuracy* | Task Background Reflectance (per cent) | D | E | F | G** | H** | I** |
| Under 40 | NI | Over 70 | 200 | 500 | 1000 | 2000 | 5000 | 10000 |
|  |  | 30–70 | 200 | 500 | 1000 | 2000 | 5000 | 10000 |
|  |  | Under 30 | 300 | 750 | 1500 | 3000 | 7500 | 15000 |
|  | I | Over 70 | 200 | 500 | 1000 | 2000 | 5000 | 10000 |
|  |  | 30–70 | 300 | 750 | 1500 | 3000 | 7500 | 15000 |
|  |  | Under 30 | 300 | 750 | 1500 | 3000 | 7500 | 15000 |
|  | C | Over 70 | 300 | 750 | 1500 | 3000 | 7500 | 15000 |
|  |  | 30–70 | 300 | 750 | 1500 | 3000 | 7500 | 15000 |
|  |  | Under 30 | 300 | 750 | 1500 | 3000 | 7500 | 15000 |
| 40–55 | NI | Over 70 | 200 | 500 | 1000 | 2000 | 5000 | 10000 |
|  |  | 30–70 | 300 | 750 | 1500 | 3000 | 7500 | 15000 |
|  |  | Under 30 | 300 | 750 | 1500 | 3000 | 7500 | 15000 |
|  | I | Over 70 | 300 | 750 | 1500 | 3000 | 7500 | 15000 |
|  |  | 30–70 | 300 | 750 | 1500 | 3000 | 7500 | 15000 |
|  |  | Under 30 | 300 | 750 | 1500 | 3000 | 7500 | 15000 |
|  | C | Over 70 | 300 | 750 | 1500 | 3000 | 7500 | 15000 |
|  |  | 30–70 | 300 | 750 | 1500 | 3000 | 7500 | 15000 |
|  |  | Under 30 | 500 | 1000 | 2000 | 5000 | 10000 | 20000 |
| Over 55 | NI | Over 70 | 300 | 750 | 1500 | 3000 | 7500 | 15000 |
|  |  | 30–70 | 300 | 750 | 1500 | 3000 | 7500 | 15000 |
|  |  | Under 30 | 300 | 750 | 1500 | 3000 | 7500 | 15000 |
|  | I | Over 70 | 300 | 750 | 1500 | 3000 | 7500 | 15000 |
|  |  | 30–70 | 300 | 750 | 1500 | 3000 | 7500 | 15000 |
|  |  | Under 30 | 500 | 1000 | 2000 | 5000 | 10000 | 20000 |
|  | C | Over 70 | 300 | 750 | 1500 | 3000 | 7500 | 15000 |
|  |  | 30–70 | 500 | 1000 | 2000 | 5000 | 10000 | 20000 |
|  |  | Under 30 | 500 | 1000 | 2000 | 5000 | 10000 | 20000 |

\* NI = not important, I – important, and C = critical
\*\* Obtained by a combination of general and supplementary lighting.

**Illuminance Values, Maintained, in Lux, for a Combination of Illuminance Categories and User, Room and Task Characteristics (For Illuminance in Footcandles, Divide by 10)**

**Fig 2-16**

## LIGHTING THE TASK

Visibility is the key to increased production. More work with fewer errors results when the task can be seen quickly, easily, and accurately. Lighting for a task should be of sufficient quantity, relatively even over the task area. For office tasks and reading and writing, it should be shadow free. Illumination on a task may be provided by an overhead system that distributes even general lighting throughout a space or by luminaires located closer to the task (local lighting) that are designed to light primarily the task area. The latter is usually what is meant by TASK LIGHTING in office applications.

| Area | Recommended Minimum Footcandles |
|---|---|
| **Industrial** | |
| **Assembly** | |
| Rough easy seeing | 30 |
| Medium | 100 |
| Fine | 500 |
| **Exterior Areas** | |
| Entrances | |
|    Active (pedestrian and/or conveyance) | 5 |
|    Inactive (normally locked, infrequently used) | 1 |
| Building surrounds | 1 |
| Active shipping area | 5 |
| Storage areas - active | 20 |
| Storage areas - inactive | 1 |
| Loading and unloading platforms | 20 |
| **Stairways, Corridors, and Other Service Areas** | 20 |
| **Storage Rooms or Warehouses** | |
| Inactive | 5 |
| Active | |
|    Rough bulky | 10 |

| | |
|---|---|
| Fine | 50 |
| **Toilets and Washrooms** | **30** |
| **Stores, Offices and Institutions** | |
| **Art Galleries** | |
| General | 30 |
| On paintings (supplementary) | 30 |
| On statuary | 100 |
| **Auditoriums** | |
| Assembly only | 15 |
|     General | 10 |
| Corridors, elevators, and stairs | 20 |
| Entrance foyer | 30 |
| Linen room | |
|     Sewing | 100 |
|     General | 20 |
| Lobby | |
|     General lighting | 10 |
|     Reading and working areas | 30 |
| **Libraries** | |
| Reading rooms and carrels | 70* |
| Stacks | 30 |
| Book repair and binding | 70 |
| Check-out catalogs | 70* |
| Card files | 100* |
| **Offices** | |
| General | |
|     Cartography, designing, detailed drafting | 200* |
|     Accounting, auditing, tabulating | |
|     Bookkeeping, business machine operation | 150* |
| Regular Office Work | |
|     Good copy | 70* |
| Regular office work— reading, transcribing, active filing, mail sorting, etc., fair-quality copy | 100* |
| Corridors, elevators, escalators, stairways (Or, not less than 1/5 the level in adjacent areas.) | 20 |

**Restaurants**
Dining Areas
    Cashier                           50
    Intimate environment     10
    Subdued environment    3
    Leisure environment      30

**Theaters**
Auditoriums
    During intermission       5
    During performance
    or presentation            0.1
Foyer
Entrance lobby

**Garages - Automobile and Truck**
Service garages
    Repairs                     100
    Active traffic areas       20
Parking garages
    Entrances                 50
    Traffic lanes             10
    Storage                  5

**Inspection**
Ordinary                     50
Difficult                      100
Highly difficult          200
Most difficult            1000

**Materials Handling**
Wrapping, packing, labeling    50
Picking stock, classifying      30
Loading, trucking           20

**Banks (see also Offices)**
Lobby                        50
Writing areas in lobby      70*

**Barber and Beauty Shops**    100

**Churches & Synagogues**
Altar, arc                  100
Pews                         15
Pulpit (supplementary)      50

**Courtrooms**

| | |
|---|---|
| Seating area | 30 |
| Court activity area | 70* |

**Hotels and Motels**
Bars and cocktail lounges
(see Restaurants)
Bathrooms

| | |
|---|---|
| General | 10 |
| Mirror | 30 |

Bedrooms

| | |
|---|---|
| Reading (books, magazines, newspapers) | 30 |
| Subdued environment | 15 |

Quick service type

| | |
|---|---|
| Bright surroundings | 100 |
| Normal surroundings | 50 |

NOTE: Footcandle levels in dining areas are highly variable. Variations depend on such factors as time of day, desired atmosphere, individuality, and attractiveness.

**Stores**
Store interiors

| | |
|---|---|
| Circulation areas | 30 |

Merchandising areas

| | |
|---|---|
| Service stores | 100 |
| Self-service stores | 200 |

Showcases and wall cases

| | |
|---|---|
| Service stores | 200 |
| Self-service stores | 500 |

Feature displays

| | |
|---|---|
| Service stores | 500 |
| Self-service stores | 1000 |
| Stockrooms | 30 |

*Equivalent sphere illumination (ESI footcandle)

**Lighting level chart**

## GLARE

Glare is not generally considered a design element or technique but rather something to be avoided. It is included as a topic in this section because the elimination of glare should be the aim of every good lighting design and it must be kept in mind in every design decision make. There are two principal types of glare to consider, Reflected Glare and Direct Glare. Reflected glare is a result of light being reflected off a surface and interfering with the task. Direct glare is the result of a light source shining directly into the observer's eyes and causing the eyes to inappropriately adjust for light levels. Direct glare causes eye strain.

## REFLECTED GLARE

### VEILING REFLECTIONS

Luminaires should be located to minimize veiling reflections. Luminaires located above and in front of a task are said to be in the OFFENDING ZONE and cause veiling reflections which reduce contrast and therefore visibility. Light falling on a task from the side or from behind the viewer eliminates veiling reflections as shown in the following figure.

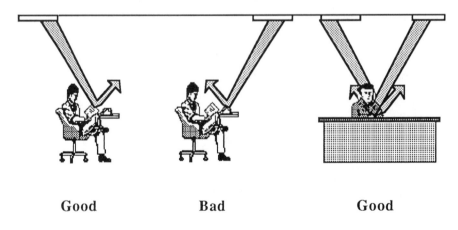

Good      Bad      Good

"Locating Light." Fig 2-17

## "BATWING" DISTRIBUTION

A broad light distribution that concentrates most of the light output in the 30 degree to 60 degree zone is known as "BATWING" DISTRIBUTION. Only a small portion of light distribution occurs in the 0 degree - 30 degree veiling reflection zone or in the 60 degree-90 degree glare zone as shown in the following figure.

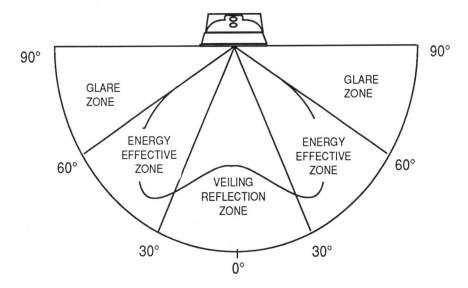

**Batwing. Fig 2-18**

Luminaires with a Batwing Distribution are thus effective in reducing both veiling reflections and glare. This is true even if the luminaires are not located in a particular relation to the task since most of the light reaching the task will come from the side rather than from in front.

## DIRECT GLARE

Direct Glare is glare caused by placing a bright source directly in the line of sight of the user of the space. The only way to eliminate direct glare is through careful placement of luminaires and by shielding. Parabolic louvers on fluorescent fixtures is an example of shielding. Direct glare is one of the most distracting forms of glare and is the most easily eliminated through careful planning.

## LOCAL LIGHTING

Local lighting refers to a luminaire located below ceiling level and/or one that is specifically related to a limited area in which visual tasks are performed. Under cabinet lighting in a kitchen, suspended downlights over a work table or a directional source aimed at a sewing machine needle all provide localized task lighting. The level of illumination increases the closer a light source is to a task. The location of local task lighting determines the presence or absence of veiling reflections. Local lighting is frequently in the line of sight, so luminaire brightness becomes particularly critical.

## OPEN PLAN TASK LIGHTING

Open plan offices consisting of individual work stations surrounded by screens and overhanging shelves and cabinets reduce the effectiveness of an overhead system of general lighting. The screens and cabinets cause shadows on the work surface by preventing much of the light from reaching the surface. Local lighting closer to the work area is needed to eliminate these shadows.

## LOCATION OF TASK LIGHTING

Lighting from one or both sides of the task avoids veiling reflections; however, luminaires placed at the sides of a task area tend to produce uneven illumination with disturbing shadows from writing utensils, books, machines, etc. A linear luminaire mounted above a work surface supplies a broad, even shadowless light; however, without a control device, it will produce severe veiling reflections because the perpendicular rays from the light source reflect from the task into the viewer's eyes and obscure the contrast necessary to see details on the task. Changing one's position or the position of the task may reduce or eliminate the problem, but relocation is not always possible.

## CONTROLLED TASK LIGHTING

With a task light in front of a worker, veiling reflections are best controlled by eliminating that portion of the light which arrives from directly ahead of the line of sight. This can be done with a lens that blocks the flow in the perpendicular direction and redirects much of the light into bilateral rays in a pattern similar to the "batwing" distribution form ceiling luminaires as shown here for a Taskline luminaire.

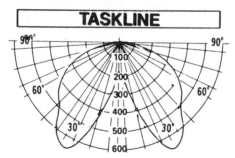

**Photometric curve, horizontal plane showing sidelighting distribution.**

Fig 2-19

Because the lighting is essentially from the sides, veiling reflections are almost eliminated, thereby increasing the effectiveness of the illumination.

## SUMMARY

The techniques of lighting design are ultimately derived from the experience of the designer. I can't emphasize often enough the need for the neophite practitioner to learn to observe intelligently every lighting installation encountered, both bad and good. Analize the technical and psychological natures of the design. Try to understand the intention of the designer. Note whether you think they succeed in the installation. Decide what you would do differently. Keep a notebook with sketches and analysis. This document will become your design bible during the start-up of your career.

# Chapter 3

## LIGHTING ENGINEERING

There is an inherent tension between the lighting designer and the electrical engineer who typically is responsible for the specifications of the design generated by the lighting aesthetician. This is unfortunate, for there need be only cooperation. This chapter is designed to give the non-engineer a bag of technical tools sufficient to keep him/her out of trouble and bridge the gap between the design and technical aspects of a project.

It is also intended to give the non-engineer an introduction to the complexities and value of the analytical process in developing and realizing the aesthetic concepts inherent in exceptional architectural lighting.

There really is no dichotomy between the two. Frank Lloyd Wright was one of the most innovative and brilliant architect of our century, and at the same time one of the most advanced structural engineers. Without the blending of the two, it is doubtful that the extent of his greatness would have been realized. This is the essence of excellence in lighting design, a blend of the aesthetic and analytical.

## CALCULATIONS

## DESIGN CONSIDERATIONS

Quantity, quality, economy, aesthetics and psychology must all be considered in designing a lighting installation. The quantity of light is the simplest factor because it is objective. The amount of light needed for a variety of tasks and environments has been determined by research and published by the Illuminating Engineering Society. The amounts are now given with weighting factors to compensate for the difficulty of the task, duration of activity, age of the viewer and reflectance of task background. While these are subjective evaluations, they still fall within specific numerical ranges. (See the IES Handbook for recommended levels of illumination).

The way to accurately determine the lighting level resulting from any choice of luminaire is to use calculations. These can be accomplished by using look-up charts provided by manufacturers or equipment and lamps, by testing an calculation using standard engineering techniques and by computer modeling of the proposed lighting environment. The end result can be expressed in charts and lists or graphically. In this chapter we will explore several important types.

Let me redefine a few of the lighting calculation terms I will be referring to in the following chapter.

The first is Zonal-Cavity calculation, sometimes referred to as the Lumen Method. This procedure takes into consideration the size and surface condition of a room, as well as the photometric and mounting characteristics of a luminaire, and provides the designer with the estimated average lighting level at a user defined worksurface. Consideration is given for lamp and other depreciation factors. This is the most basic lighting calculation. It also is so generic as to be of minimal use for any but the simplest layout.

The second is VCP or Visual Comfort Probability. VCP is defined as the percent of observers with normal vision who will be comfortable in that specific visual environment. A VCP of 70 or more indicates there little direct glare or comfort problem with the design. This can be useful in evaluating the quality of lighting in a room.

Third, Point-to-Point calculations represent the lighting level to be expected at various points throughout a lighting environment expressed on various planes such as wall surfaces or work planes. This calculation represents lighting quantity distribution in the design.

Finally, ESI or Equivalent Spherical Illumination and related factors, CRF or Contrast Rendition Factor, and LEF or Lighting Effectiveness Factor. The ESI/CRF calculation is also concerned with quality, and evaluates the lighting system with reference to seeing ability, defined as degree of contrast for a given task, in a standard environment, and compared to the actual defined design being evaluated. CRF/ESI analysis quantifies a qualitative prejudgment in the design process and is a limited but powerful evaluation tool.

## POINT METHOD FOR ILLUMINATION-AT-A-POINT FOR SINGLE LUMINAIRES.

### HOW TO CALCULATE FOOTCANDLES AT A POINT

The basis of lighting calculations is to determine how many light source lumens arrive on a given surface. The two commonly used methods of calculating footcandles are the SINGLE UNIT POINT METHOD for a specific limited area and the ZONAL CAVITY METHOD for average general illumination in a space. In this lesson we treat only the first of these.

THE POINT METHOD is used to predict the illumination from a single source on a relatively small target using a candlepower distribution curve. The formulas are accurate only when the largest dimension of the source is less than one-fifth the distance from the source to the work-plane. The calculations do not take into consideration any reflected light from nearby surfaces.

For example:

PROBLEM -- Find the footcandle level on a surface directly under a light source aimed straight down as in Fig. 3-1A.

SOLUTION -- From a candlepower distribution curve determine the candelas (I) at 0° (known as the normal angle), then divide by the distance (D) squared as in fc=1.

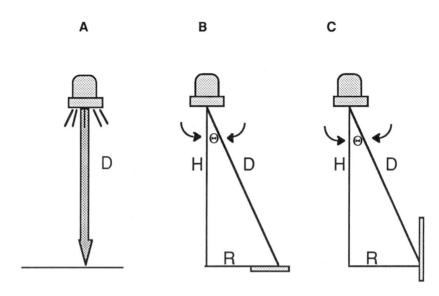

Illumination at a point from a point source. Fig 3-1

Another factor must be used to determine the amount of light arriving on a surface from an angle other than normal (0°), that is, the cosine of the angle, which can be found in any cosine table.

PROBLEM -- Find the initial footcandles arriving on a HORIZONTAL SURFACE at a given angle from a source aimed straight down but located to one side of it as in Fig. 3-1B

SOLUTION -- From the candle power distribution curve, find the candelas (I) at the desired angle, then multiple by the COSINE of the angle (cos 0) found in the cosine table, then divide by the square of the distance (D).

$$fc = \frac{(I) \times (\cos \Theta)}{D^2}$$

(Note that $D^2$ is equal to $H^2 + R^2$)

## Chapter Three  Lighting Engineering  75

If the target area is on a VERTICAL SURFACE as in Fig. 3-1C, use the SINE of the angle (sin 0) to determine footcandles. The formula is:

$$fc = \frac{(I) \times (\sin \Theta)}{D^2}$$

## CALCULATOR CHARTS FOR SINGLE LUMINAIRES

Manufacturers brochures include charts that can be used to estimate lighting levels quickly without lengthy calculations.

SINGLE UNIT  CAT. NO. 7054

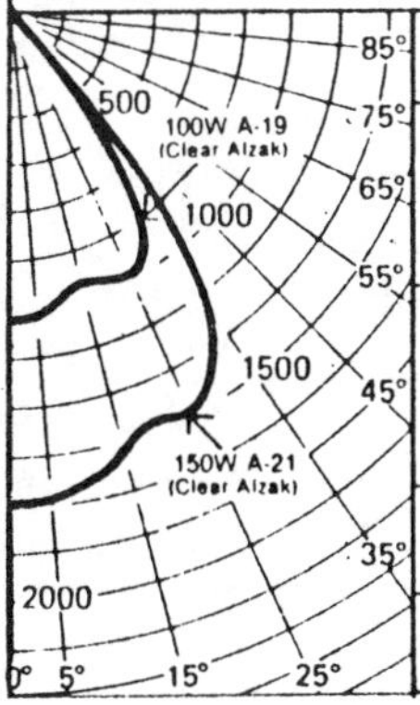

150W A-21 (Clear Alzak)

| Height to Floor | Foot-candles | Beam Dia. † |
|---|---|---|
| | At 30" above floor | |
| 8' | 55 | 11' |
| 10 | 29 | 15' |
| 12' | 18 | 19' |
| 14' | 13 | 23' |

Beam Spread † = 90°

†At 10% of Maximum Candlepower

**Typical data for a single incandescent luminaire. Fig 3-2**

This figure shows a candlepower distribution curve and its translation into predicted illumination from a single luminaire at various heights, as well as the width of the beam at 10% of the maximum candlepower. From this chart you can learn that the 7054 open reflector downlight with a 150A21 lamp delivers approximately 55 footcandles directly underneath it on a work surface 30" above the floor when mounted in an 8' high ceiling and the beam diameter will be 11'.

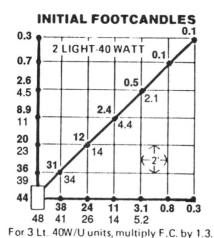

**Typical data for a single fluorescent luminaire. Fig 3-3**

This figure shows illumination at specific points from a single fluorescent luminaire -- a 2' x 2', two lamp, Parabolic Louver unit. Initial footcandle values under a single luminaire are shown on a 2' x 2', grid. Two values are given; one for a luminaire in a small room with reflectances of ceiling: 80%, walls: 50%, and floors: 20%; and the other, in the middle of a large room where no reflection from the walls contributes to the task lighting, hence effective reflectances are; ceiling: 80%, walls: 0%, and floor: 20%. From the chart we can determine that at a point 4' in front of and 4' to right of center, there will be approximately 12 fc in a large room and 14 fc in a small room. Only one quadrant of the light pattern is shown since the values are the same for the other three quadrants.

With the increasing use of non-uniform lighting in offices, information on task illumination from several luminaires is required. Footcandle levels at a specific point under a group of luminaires may be determined by super-imposing the graphs for all luminaires and adding up the contribution from each.

Chapter Three    Lighting Engineering    77

A LAMP SELECTOR GUIDE (see Fig 3-4) shows the amount and spread of light at various distances for lamps at 0° aiming angle. The beam spreads indicated are to 10% of maximum candlepower. The LIGHTING PERFORMANCE DATA, Fig. 3-4, shows footcandle charts and the data needed to develop lighting performance information for angled luminaires, singly or in multiple installations. The beam length (L) and beam width (W) are to the point where the candlepower drops to 10% of maximum.

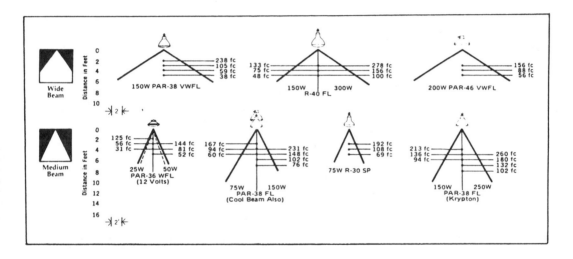

**Typical data from Lighting Performance Data of Lytespan Cagalog.**

**Fig 3-4**

## ZONAL CAVITY METHOD FOR AVERAGE ILLUMINATION

### ZONAL CAVITY METHOD

Rather than calculating every point in a room, adding them together and adjusting for inter-reflection to arrive at AVERAGE GENERAL ILLUMINATION, we use the ZONAL CAVITY METHOD. This method assumes an area to be lighted to be made up of a series of zones or cavities.

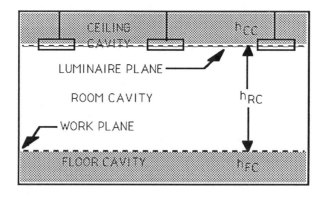

**A room is divided into three cavities. Fig 3-5**

The Zonal Cavity Method takes the size and shape of a space and the reflectance values of all surfaces as well as the luminaire CU into consideration in computing the amount of light delivered to the work plane from an overhead lighting system. In general, the higher and narrower the room, the larger the percentage of light that will be absorbed by the walls and the lower the efficiency of the system.

Chapter Three    Lighting Engineering

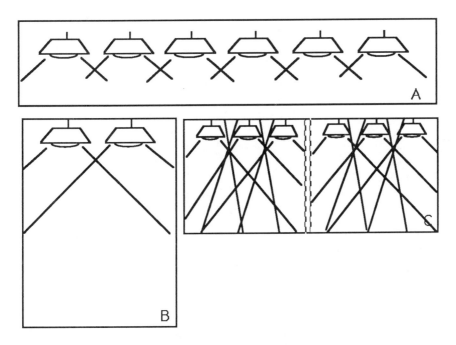

**Light is absorbed. Fig 3-6**

## METHOD PROCEDURE

To determine illumination levels from a specific layout of luminaires with the Zonal Cavity Method, use the following procedure:

1. Determine the CAVITY RATIOS either by referring to a Cavity Ratio Table such as Figure 13-4 at the end of this lesson, or in APPLICATION GUIDE 10.01, or by using these formulas:

$$\text{Ceiling Cavity Ratio (CCR)} = \frac{5h_{CC} \times (L + W)}{L + W}$$

$$\text{Room Cavity Ratio (RCR)} = \frac{5h_{RC} \times (L + W)}{L + W}$$

$$\text{Floor Cavity Ratio (FCR)} = \frac{5h_{FC} \times (L + W)}{L + W}$$

where h is height in feet, L is the length of the room in feet and W is the width of the room in feet. With recessed or surface equipment, the ceiling cavity is 0 and need not be considered.

2. Determine the EFFECTIVE CEILING CAVITY REFLECTANCE and EFFECTIVE FLOOR CAVITY REFLECTANCE, i.e., the percent of light reflected from the surfaces compared to the amount of light striking them. See Table II, Fig 3-12. Flat white paint has a reflectance of 80-85%, while beige or light green may only reflect 50-60% of the incident light. The higher the reflectances, the more useful light reaches the work plane. The floor cavity reflectance is assumed to be 20% in Quick Calculator Charts. Where there is a substantial difference in the floor reflectance, see Table III, Fig 3-13

3. Select the COEFFICIENT OF UTILIZATION (CU) from the table supplied for the luminaire on the specification sheet by finding the Room Cavity Ratio in the left column, then reading across to the appropriate reflectance column (See Fig 3-7. If the RCR is a fraction, interpolate for the CU.

| | | % Effective Ceiling Cavity Reflectance | | | | | | | |
|---|---|---|---|---|---|---|---|---|---|
| 4 LIGHT - 40 WATT | | | | | I.T.L. Report No. 16466C | | | | |
| | | 80 | | | 50 | | | 10 | |
| | | % Wall Reflectance | | | | | | | |
| | | 50 | 30 | 10 | 50 | 30 | 10 | 50 | 30 | 10 |
| Room Cavity Ratio | 1 | .63 | .62 | .60 | .60 | .58 | .57 | .56 | .55 | .54 |
| | 2 | .58 | .55 | .53 | .55 | .53 | .51 | .52 | .50 | .49 |
| | 3 | .53 | .49 | .47 | .50 | .48 | .45 | .48 | .48 | .44 |
| | 4 | .48 | .44 | .41 | .46 | .43 | .40 | .44 | .41 | .39 |
| | 5 | .43 | .39 | .36 | .42 | .38 | .36 | .40 | .37 | .35 |
| | 6 | .40 | .35 | .32 | .38 | .34 | .32 | .36 | .33 | .31 |
| | 7 | .36 | .31 | .28 | .34 | .31 | .28 | .33 | .30 | .28 |
| | 8 | .32 | .28 | .25 | .31 | .27 | .24 | .30 | .26 | .24 |
| | 9 | .29 | .24 | .21 | .28 | .24 | .21 | .27 | .23 | .21 |
| | 10 | .26 | .22 | .19 | .25 | .21 | .19 | .24 | .21 | .19 |

For 3 Lt.-40 Watt units, multiply C.U.'s by 1.1. SPACING RATIO:
For 6 Lt.-40 Watt units, multiply C.U.'s by 0.9. PARALLEL = 1.2
                                              PERPENDICULAR = 1.4

**CU table for a 4 Lt.-40W luminaire. Fig 3-7**

4. Determine the MAINTENANCE FACTOR (MF). Since lamps, luminaires and reflecting surfaces all deteriorate over time, light loss must be factored into calculations to determine realistic maintained levels of illumination. RECOVERABLE FACTORS such as Room Surface Dirt Depreciation (RSDD), Lamp Lumen Depreciation (RSDD), Lamp Lumen Depreciation (LLD), Luminaire Dirt Depreciation (LDD) and lamp burn outs are rated numerically for calculations. RSDD and LDD are shown in Figures 3-8 and 3-9.

| | Luminaire Distribution Type | | | | | | | | | | | | | | | | | | | |
|---|---|---|---|---|---|---|---|---|---|---|---|---|---|---|---|---|---|---|---|---|
| | Direct | | | | Semi-Direct | | | | Direct-Indirect | | | | Semi-Indirect | | | | Indirect | | | |
| Per cent expected Dirt Depreciation | 10 | 20 | 30 | 40 | 10 | 20 | 30 | 40 | 10 | 20 | 30 | 40 | 10 | 20 | 30 | 40 | 10 | 20 | 30 | 40 |
| Room Cavity Ratio | | | | | | | | | | | | | | | | | | | | | |
| 1 | 98 | 96 | 94 | 92 | 97 | 92 | 89 | 84 | 94 | 87 | 80 | 76 | 94 | 87 | 80 | 73 | 90 | 80 | 70 | 60 |
| 2 | 98 | 96 | 94 | 92 | 96 | 92 | 88 | 83 | 94 | 87 | 80 | 75 | 94 | 87 | 79 | 72 | 90 | 80 | 69 | 59 |
| 3 | 98 | 95 | 93 | 90 | 96 | 91 | 87 | 82 | 94 | 86 | 79 | 74 | 94 | 86 | 78 | 71 | 90 | 79 | 68 | 58 |
| 4 | 97 | 95 | 92 | 90 | 95 | 90 | 85 | 80 | 94 | 86 | 79 | 73 | 94 | 86 | 78 | 70 | 89 | 78 | 67 | 56 |
| 5 | 97 | 94 | 91 | 89 | 94 | 90 | 84 | 79 | 93 | 86 | 78 | 72 | 93 | 86 | 77 | 69 | 89 | 78 | 66 | 55 |
| 6 | 97 | 94 | 91 | 88 | 94 | 89 | 83 | 78 | 93 | 85 | 78 | 71 | 93 | 85 | 76 | 68 | 89 | 77 | 66 | 54 |
| 7 | 97 | 94 | 90 | 87 | 93 | 88 | 82 | 77 | 93 | 84 | 77 | 70 | 93 | 84 | 76 | 68 | 89 | 76 | 65 | 53 |
| 8 | 96 | 93 | 89 | 86 | 93 | 87 | 81 | 75 | 93 | 84 | 76 | 69 | 93 | 84 | 76 | 68 | 88 | 76 | 64 | 52 |
| 9 | 96 | 92 | 88 | 85 | 93 | 87 | 80 | 74 | 93 | 84 | 76 | 68 | 93 | 84 | 75 | 67 | 88 | 75 | 63 | 51 |
| 10 | 96 | 92 | 87 | 83 | 93 | 86 | 79 | 72 | 93 | 84 | 75 | 67 | 92 | 83 | 75 | 67 | 88 | 75 | 62 | 50 |

**Room Surface Dirt Depreciation Factors. Fig 3-8**

**82**  *Architectural Lighting Design*

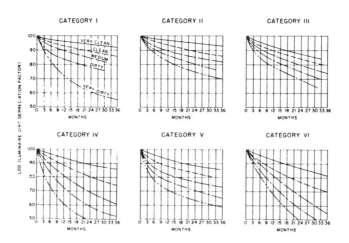

**Luminaire Dirt Depreciation Factors. Fig 3-9**

LLD and lamp burn out factors may be obtained from manufacturers' catalogs. The total light loss or Maintenance Factor (MF) is the product of multiplying all these contributing factors together. To speed calculations, a MF is frequently shown in the various luminaire catalogs.

5. Compute the AVERAGE FOOTCANDLE level as follows:

$$\text{footcandles} = \frac{\text{no. of luminaires} \times \text{no. of lamps per luminaire} \times \text{initial lumens per lamp} \times \text{Coefficient of Utilization} \times \text{Maintenance Factor}}{\text{area in square feet}}$$

For example, in an office that is 24' x 70' with a 9'6" ceiling height, 24 recessed 2' x 4' luminaires with four 40 watt fluorescent lamps (3200 L each) are installed in three rows of eight each. The ceiling is 80% reflectance; the walls, 50% reflectance; the floor, 20% reflectance. To find how many footcandles the lighting system delivers on the working plane follow the above procedure.

1. Determine the Room Cavity Ratio (RCR) from Figure 3-11. Locate width at 24 and length at 70. Read across to the line under 7 (9.5' ceiling height minus 2.5' to the work surface) to find the RCR of 2.

## Chapter Three  Lighting Engineering

2. Go the the CU table in Figure 13-6. Find the RCR of 2 in the left hand column, read across to the line under 80/50 (ceiling and wall reflectances) to find the CU = .58.

3. Assume a Maintenance Factor of .70

4. Work formula for average maintained footcandles, i.e.

$$fc = \frac{\text{no. of luminaires (24)} \times \text{no. lamps per luminaire (4)} \times \text{initial lumens per lamp (3200)} \times \text{CU (.58)} \times \text{MF (.70)}}{\text{area (24' } \times \text{ 70')}}$$

$$= \frac{24 \times 4 \times 3200 \times .58 \times .70}{1680}$$

$$= 74$$

5. To find HOW MANY LUMINAIRES are needed to achieve a certain footcandle level, the formula can be rewritten as:

$$\text{Number of Luminaires} = \frac{\text{area in square feet} \times \text{footcandles (maintained)}}{\text{lamps per luminaire} \times \text{initial lumens per lamp} \times \text{coefficient of utilization} \times \text{maintenance factor}}$$

$$= \frac{(24 \times 70) \times 74}{4 \times 3200 \times .58 \times .70}$$

$$= 24$$

# Architectural Lighting Design

## Average Illumination Calculation Sheet

### GENERAL INFORMATION

Project Identification:_____
(Give name of area and/or building and room number)

Average maintained illumination for design:____footcandles

Luminaire data:

Manufacturer:_____

Catalog number:_____

Lamp data

Type and Color:_____

Number per luminaire:____

Total lumens per luminaire_____

### SELECTION OF COEFFICIENT OF UTILIZATION

Step 1: Fill in sketch at right

Step 2: Determine Cavity Ratios from Table. or by formulas.

Room Cavity Ratio, RCR = _____

Ceiling Cavity Ratio, CCR = _____

Floor Cavity Ratio, FCR = _____

Step 3: Obtain Effective Ceiling Cavity Reflectance from Table.  $\rho CC$ = ____

Step 4: Obtain Effective Floor Cavity Reflectance from Table  $\rho FC$ = ____

Step 5: Obtain Coefficient of Utilization (CU) from Manufacturer's Data.  CU = ____

### SELECTION OF LIGHT LOSS FACTORS

| Unrecoverable | | Recoverable | |
|---|---|---|---|
| Luminaire ambient temperature | | Room Surface dirt depreciation RSDD | _____ |
| Voltage to luminaire | _____ | Lamp lumen depreciation LLD | _____ |
| Ballast factor | _____ | Lamp burnouts factor LBO | _____ |
| Luminaire surface depreciation | _____ | Luminaire dirt depreciation LDD | _____ |

Total light loss factor, LLF (product of individual factors above):_____

### CALCULATIONS

(Average Maintained Illumination Level)

$$\text{Number of Luminaires} = \frac{\text{(Footcandles) X (Area in Square feet)}}{\text{(Lumens per Luminaire) X (CU) X (LLF)}}$$

= _____ =

$$\text{Footcandles} = \frac{\text{(Number of Luminaires) X (Lumens per Luminaire) X (CU) X (LLF)}}{\text{(Area in Square Feet)}}$$

= _____ =

Calculated by:_____ Date:_____

## QUICK CALCULATOR CHARTS

For quick evaluation of luminaire performance, use a Quick Calculator chart (as shown in Fig. 3-10) to estimate average INITIAL footcandles and the effect on illumination levels of various spacings of the luminaires. For example, the chart for a four lamp luminaire shows that approximately 100 fc's are delivered initially when located 9' on center in a room with an RCR of 1. (For maintained fc approximation, multiply by a MF). Or we determine that with an RCR of 5, for 75 fc initial, the luminaires should be approximately 8.5' apart. To compensate for variables such as different types or numbers of lamps or reflector finishes, multiply by the appropriate correction factor shown with the chart. For instance, if it were a three lamp unit used in the last example, multiply the 8.5' spacing by .8 for an approximate spacing of 6.8' on center. The actual spacing will in most cases be dictated by some modular element in the building such as a ceiling tile. Thus the actual spacings in these examples might be 8' and 6'.

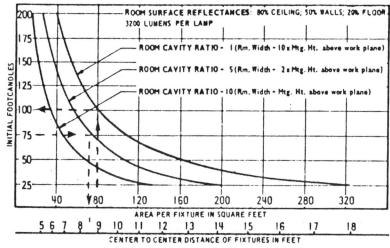

For 3 Lt.-40W units, multiply chart values by 0.8.
For 6 Lt.-40W units, multiply chart values by 1.4

**Quick Calculator Chart for a 4 Lt.-40W luminaire. Fig 3-10**

# 86 Architectural Lighting Design

Cavity Ratios Chart Table I. Fig 3-11

TABLE II – Effective Ceiling or Floor Cavity Reflectance for Various Reflectance Combinations

| PER CENT CEILING OR FLOOR REFLECTANCE | | 90 | | | | 80 | | | | 70 | | | 50 | | | 30 | | | | 10 | | |
|---|---|---|---|---|---|---|---|---|---|---|---|---|---|---|---|---|---|---|---|---|---|---|
| PER CENT WALL REFLECTANCE | | 90 | 70 | 50 | 30 | 80 | 70 | 50 | 30 | 70 | 50 | 30 | 70 | 50 | 30 | 65 | 50 | 30 | 10 | 50 | 30 | 10 |
| Ceiling or Floor Cavity Ratio | 0 | 90 | 90 | 90 | 90 | 80 | 80 | 80 | 80 | 70 | 70 | 70 | 50 | 50 | 50 | 30 | 30 | 30 | 30 | 10 | 10 | 10 |
| | 0.1 | 90 | 89 | 88 | 87 | 79 | 79 | 78 | 78 | 69 | 69 | 68 | 59 | 49 | 48 | 30 | 30 | 29 | 29 | 10 | 10 | 10 |
| | 0.2 | 89 | 88 | 86 | 85 | 79 | 78 | 77 | 76 | 68 | 67 | 66 | 49 | 48 | 47 | 30 | 29 | 29 | 28 | 10 | 10 | 9 |
| | 0.3 | 89 | 87 | 85 | 83 | 78 | 77 | 75 | 74 | 68 | 66 | 64 | 49 | 47 | 46 | 30 | 29 | 28 | 27 | 10 | 10 | 9 |
| | 0.4 | 88 | 86 | 83 | 81 | 78 | 76 | 74 | 72 | 67 | 65 | 63 | 48 | 46 | 45 | 30 | 29 | 27 | 26 | 11 | 10 | 9 |
| | 0.5 | 88 | 85 | 81 | 78 | 77 | 75 | 73 | 70 | 66 | 64 | 61 | 48 | 46 | 44 | 29 | 28 | 27 | 25 | 11 | 10 | 9 |
| | 0.6 | 88 | 84 | 80 | 76 | 77 | 75 | 71 | 68 | 65 | 62 | 59 | 47 | 45 | 43 | 29 | 28 | 26 | 25 | 11 | 10 | 9 |
| | 0.7 | 88 | 83 | 78 | 74 | 76 | 74 | 70 | 66 | 65 | 61 | 58 | 47 | 44 | 42 | 29 | 28 | 26 | 24 | 11 | 10 | 8 |
| | 0.8 | 87 | 82 | 77 | 73 | 75 | 73 | 69 | 65 | 64 | 60 | 56 | 47 | 43 | 41 | 29 | 27 | 25 | 23 | 11 | 10 | 8 |
| | 0.9 | 87 | 81 | 76 | 71 | 75 | 72 | 68 | 63 | 63 | 59 | 55 | 46 | 43 | 40 | 29 | 27 | 25 | 22 | 11 | 9 | 8 |
| | 1.0 | 86 | 80 | 74 | 69 | 74 | 71 | 66 | 61 | 63 | 58 | 53 | 46 | 42 | 39 | 29 | 27 | 24 | 22 | 11 | 9 | 8 |
| | 1.1 | 86 | 79 | 73 | 67 | 74 | 71 | 65 | 60 | 62 | 57 | 52 | 46 | 41 | 38 | 29 | 26 | 24 | 21 | 11 | 9 | 8 |
| | 1.2 | 86 | 78 | 72 | 65 | 73 | 70 | 64 | 58 | 61 | 56 | 50 | 45 | 41 | 37 | 29 | 26 | 23 | 20 | 12 | 9 | 7 |
| | 1.3 | 85 | 78 | 70 | 64 | 73 | 69 | 63 | 57 | 61 | 55 | 49 | 45 | 40 | 36 | 29 | 26 | 23 | 20 | 12 | 9 | 7 |
| | 1.4 | 85 | 77 | 69 | 62 | 72 | 68 | 62 | 55 | 60 | 54 | 48 | 45 | 40 | 35 | 28 | 26 | 22 | 19 | 12 | 9 | 7 |
| | 1.5 | 85 | 76 | 68 | 61 | 72 | 68 | 61 | 54 | 59 | 53 | 47 | 44 | 39 | 34 | 28 | 25 | 22 | 18 | 12 | 9 | 7 |
| | 1.6 | 85 | 75 | 66 | 59 | 71 | 67 | 60 | 53 | 59 | 52 | 45 | 44 | 39 | 33 | 28 | 25 | 21 | 18 | 12 | 9 | 7 |
| | 1.7 | 84 | 74 | 65 | 58 | 71 | 66 | 59 | 52 | 58 | 51 | 44 | 44 | 38 | 32 | 28 | 25 | 21 | 17 | 12 | 9 | 7 |
| | 1.8 | 84 | 73 | 64 | 56 | 70 | 65 | 58 | 50 | 57 | 50 | 43 | 43 | 37 | 32 | 28 | 25 | 21 | 17 | 12 | 9 | 6 |
| | 1.9 | 84 | 73 | 63 | 55 | 70 | 65 | 57 | 49 | 57 | 49 | 42 | 43 | 37 | 31 | 28 | 25 | 20 | 16 | 12 | 9 | 6 |
| | 2.0 | 83 | 72 | 62 | 53 | 69 | 64 | 56 | 48 | 56 | 48 | 41 | 43 | 37 | 30 | 28 | 24 | 20 | 16 | 12 | 9 | 6 |
| | 2.1 | 83 | 71 | 61 | 52 | 69 | 63 | 55 | 47 | 56 | 47 | 40 | 43 | 36 | 29 | 28 | 24 | 20 | 16 | 13 | 9 | 6 |
| | 2.2 | 83 | 70 | 60 | 51 | 68 | 63 | 54 | 45 | 55 | 46 | 39 | 42 | 36 | 29 | 28 | 24 | 19 | 15 | 13 | 9 | 6 |
| | 2.3 | 83 | 69 | 59 | 50 | 68 | 62 | 53 | 44 | 54 | 46 | 38 | 42 | 35 | 28 | 28 | 24 | 19 | 15 | 13 | 9 | 6 |
| | 2.4 | 82 | 68 | 58 | 48 | 67 | 61 | 52 | 43 | 54 | 45 | 37 | 42 | 35 | 27 | 28 | 24 | 19 | 14 | 13 | 9 | 6 |
| | 2.5 | 82 | 68 | 57 | 47 | 67 | 61 | 51 | 42 | 53 | 44 | 36 | 41 | 34 | 27 | 27 | 23 | 18 | 14 | 13 | 9 | 6 |
| | 2.6 | 82 | 67 | 56 | 46 | 66 | 60 | 50 | 41 | 53 | 43 | 35 | 41 | 34 | 26 | 27 | 23 | 18 | 13 | 13 | 9 | 5 |
| | 2.7 | 82 | 66 | 55 | 45 | 66 | 60 | 49 | 40 | 52 | 43 | 34 | 41 | 33 | 26 | 27 | 23 | 18 | 13 | 13 | 9 | 5 |
| | 2.8 | 81 | 66 | 54 | 44 | 66 | 59 | 48 | 39 | 52 | 42 | 33 | 41 | 33 | 25 | 27 | 23 | 18 | 13 | 13 | 9 | 5 |
| | 2.9 | 81 | 65 | 53 | 43 | 65 | 58 | 48 | 38 | 51 | 41 | 33 | 40 | 33 | 25 | 27 | 23 | 17 | 12 | 13 | 9 | 5 |
| | 3.0 | 81 | 64 | 52 | 42 | 65 | 58 | 47 | 38 | 51 | 40 | 32 | 40 | 32 | 24 | 27 | 22 | 17 | 12 | 13 | 8 | 5 |
| | 3.1 | 80 | 64 | 51 | 41 | 64 | 57 | 46 | 37 | 50 | 40 | 31 | 40 | 32 | 24 | 27 | 22 | 17 | 12 | 13 | 8 | 5 |
| | 3.2 | 80 | 63 | 50 | 40 | 64 | 57 | 45 | 36 | 50 | 39 | 30 | 39 | 31 | 23 | 27 | 22 | 16 | 11 | 13 | 8 | 5 |
| | 3.3 | 80 | 62 | 49 | 39 | 64 | 56 | 44 | 35 | 49 | 39 | 30 | 39 | 31 | 23 | 27 | 22 | 16 | 11 | 13 | 8 | 5 |
| | 3.4 | 80 | 62 | 48 | 38 | 63 | 56 | 44 | 34 | 49 | 38 | 29 | 39 | 31 | 22 | 27 | 22 | 16 | 11 | 13 | 8 | 5 |
| | 3.5 | 79 | 61 | 48 | 37 | 63 | 55 | 43 | 33 | 48 | 38 | 29 | 39 | 30 | 22 | 26 | 22 | 16 | 11 | 13 | 8 | 5 |
| | 3.6 | 79 | 60 | 47 | 36 | 62 | 54 | 42 | 33 | 48 | 37 | 28 | 39 | 30 | 21 | 26 | 21 | 15 | 10 | 13 | 8 | 5 |
| | 3.7 | 79 | 60 | 46 | 35 | 62 | 54 | 42 | 32 | 48 | 37 | 27 | 38 | 30 | 21 | 26 | 21 | 15 | 10 | 13 | 8 | 4 |
| | 3.8 | 79 | 59 | 45 | 35 | 62 | 53 | 41 | 31 | 47 | 36 | 27 | 38 | 29 | 21 | 26 | 21 | 15 | 10 | 13 | 8 | 4 |
| | 3.9 | 78 | 59 | 45 | 34 | 61 | 53 | 40 | 30 | 47 | 35 | 26 | 38 | 29 | 20 | 26 | 21 | 15 | 10 | 13 | 8 | 4 |
| | 4.0 | 78 | 58 | 44 | 33 | 61 | 52 | 40 | 30 | 46 | 35 | 26 | 38 | 29 | 20 | 26 | 21 | 15 | 9 | 13 | 8 | 4 |
| | 4.1 | 78 | 57 | 43 | 32 | 60 | 52 | 39 | 29 | 46 | 35 | 25 | 37 | 28 | 20 | 26 | 21 | 14 | 9 | 13 | 8 | 4 |
| | 4.2 | 78 | 57 | 43 | 32 | 60 | 51 | 39 | 29 | 46 | 34 | 25 | 37 | 28 | 19 | 26 | 20 | 14 | 9 | 13 | 8 | 4 |
| | 4.3 | 78 | 56 | 42 | 31 | 60 | 51 | 38 | 28 | 45 | 34 | 25 | 37 | 28 | 19 | 26 | 20 | 14 | 9 | 13 | 8 | 4 |
| | 4.4 | 77 | 56 | 41 | 30 | 59 | 51 | 38 | 28 | 45 | 34 | 24 | 37 | 27 | 19 | 26 | 20 | 14 | 8 | 13 | 8 | 4 |
| | 4.5 | 77 | 55 | 41 | 30 | 59 | 50 | 37 | 27 | 45 | 33 | 24 | 37 | 27 | 19 | 25 | 20 | 14 | 8 | 14 | 8 | 4 |
| | 4.6 | 77 | 55 | 40 | 29 | 59 | 50 | 37 | 26 | 44 | 33 | 24 | 36 | 27 | 18 | 25 | 20 | 14 | 8 | 14 | 8 | 4 |
| | 4.7 | 77 | 54 | 40 | 29 | 58 | 49 | 36 | 26 | 44 | 33 | 23 | 36 | 26 | 18 | 25 | 20 | 13 | 8 | 14 | 8 | 4 |
| | 4.8 | 76 | 54 | 39 | 28 | 58 | 49 | 36 | 25 | 44 | 32 | 23 | 36 | 26 | 18 | 25 | 19 | 13 | 8 | 14 | 8 | 4 |
| | 4.9 | 76 | 53 | 38 | 28 | 58 | 49 | 35 | 25 | 44 | 32 | 23 | 36 | 26 | 18 | 25 | 19 | 13 | 7 | 14 | 8 | 4 |
| | 5.0 | 76 | 53 | 38 | 27 | 57 | 48 | 35 | 25 | 43 | 32 | 22 | 36 | 26 | 17 | 25 | 19 | 13 | 7 | 14 | 8 | 4 |

Table II. Fig 3-12

| % EFFECTIVE CEILING CAVITY REFLECTANCE, $\rho_{cc}$ | 80 | | | 70 | | | 50 | | | 10 | | |
|---|---|---|---|---|---|---|---|---|---|---|---|---|
| % WALL REFLECTANCE, $\rho_w$ | 50 | 30 | 10 | 50 | 30 | 10 | 50 | 30 | 10 | 50 | 30 | 10 |
| ROOM CAVITY RATIO | | | | | | | | | | | | |
| 1 | 1.08 | 1.08 | 1.07 | 1.07 | 1.06 | 1.06 | 1.05 | 1.04 | 1.04 | 1.01 | 1.01 | 1.01 |
| 2 | 1.07 | 1.06 | 1.05 | 1.06 | 1.05 | 1.04 | 1.04 | 1.03 | 1.03 | 1.01 | 1.01 | 1.01 |
| 3 | 1.05 | 1.04 | 1.03 | 1.05 | 1.04 | 1.03 | 1.03 | 1.03 | 1.02 | 1.01 | 1.01 | 1.01 |
| 4 | 1.05 | 1.03 | 1.02 | 1.04 | 1.03 | 1.02 | 1.03 | 1.02 | 1.02 | 1.01 | 1.01 | 1.00 |
| 5 | 1.04 | 1.03 | 1.02 | 1.03 | 1.02 | 1.02 | 1.02 | 1.02 | 1.01 | 1.01 | 1.01 | 1.00 |
| 6 | 1.03 | 1.02 | 1.01 | 1.03 | 1.02 | 1.01 | 1.02 | 1.02 | 1.01 | 1.01 | 1.01 | 1.00 |
| 7 | 1.03 | 1.02 | 1.01 | 1.03 | 1.02 | 1.01 | 1.02 | 1.01 | 1.01 | 1.01 | 1.01 | 1.00 |
| 8 | 1.03 | 1.02 | 1.01 | 1.02 | 1.02 | 1.01 | 1.02 | 1.01 | 1.01 | 1.01 | 1.01 | 1.00 |
| 9 | 1.02 | 1.01 | 1.01 | 1.02 | 1.01 | 1.01 | 1.02 | 1.01 | 1.01 | 1.01 | 1.01 | 1.00 |
| 10 | 1.02 | 1.01 | 1.01 | 1.02 | 1.01 | 1.01 | 1.02 | 1.01 | 1.01 | 1.01 | 1.01 | 1.00 |

**Table III Chart of Effective Floor Capacity Reflectances.**
**Fig 3-13**

## COMPUTER CALCULATIONS

Over the past ten years, computer programs have been developed that begin to address real lighting design problems in ways that can be quantified. The primary issues confronted are visual comfort and glare. Until recently, useful and economic calculation of equivalent sphere illumination and visual comfort probability have been so complex and tedious that if done at all they required access to mainframe computers and were beyond the resources allocated to an ordinary architectural project. Newer "what-if?" presentations such as extensive point calculations and perspective renderings of lighting design results were only a dream. Now most of these are available economically in any design office to run on the IBM PC or its equivalent. It is now possible to accurately model complex lighting installations in minutes instead of weeks and at costs that designers can no longer afford to ignore.

While there were over 23-PC based lighting design software suppliers listed in a recent issue of "Lighting Design and Application," a publication of the Illumination Engineering Society, I will focus on the products of Lighting Technologies Inc. The largest selling microcomputer based software for lighting design is LUMEN-MICRO which was co-developed by David L. DiLaura, author of LUMEN-II and LUMEN-II. The older LUMEN programs have been the most influential time-share computer modeling systems. The PC-based products substantially capture the power, if not the speed, of their elder brothers. LUMEN-MICRO, the complex full featured modeling system, is the premier interior lighting design product. LUMEN$, its inexpensive junior brother, provides the traditional zonal-cavity or lumen method tabulations, as well as economic analysis of comparative systems. LUMEN-POINT is the outdoor analysis software for roadway, parking lot and other installations.

# Architectural Lighting Design

| Vertical Angles | Lateral Angles | | | | | | | | | | | | | | | |
|---|---|---|---|---|---|---|---|---|---|---|---|---|---|---|---|---|
| | .0 | 5.0 | 15.0 | 25.0 | 35.0 | 45.0 | 55.0 | 65.0 | 66.0 | 75.0 | 85.0 | 90.0 | 95.0 | 105.0 | 115.0 | 125.0 |
| .0 | 1644 | 1644 | 1644 | 1644 | 1644 | 1644 | 1644 | 1644 | 1644 | 1644 | 1644 | 1644 | 1644 | 1644 | 1644 | 1644 |
| 5.0 | 1733 | 1736 | 1760 | 1811 | 1882 | 1957 | 2025 | 2084 | 2077 | 2111 | 2123 | 2125 | 2120 | 2091 | 2042 | 1962 |
| 15.0 | 1879 | 1909 | 2157 | 2471 | 2663 | 2736 | 2773 | 2780 | 2775 | 2778 | 2768 | 2763 | 2753 | 2712 | 2668 | 2605 |
| 25.0 | 2283 | 2354 | 2919 | 3301 | 3374 | 3277 | 3097 | 2953 | 2941 | 2885 | 2848 | 2838 | 2831 | 2807 | 2773 | 2963 |
| 35.0 | 2171 | 2262 | 2975 | 3340 | 3593 | 3812 | 3729 | 3330 | 3286 | 2992 | 2911 | 2904 | 2909 | 2892 | 2853 | 2782 |
| 45.0 | 2318 | 2386 | 2943 | 3284 | 3927 | 4078 | 4190 | 4314 | 4292 | 4080 | 3537 | 3493 | 3505 | 3413 | 3155 | 2907 |
| 55.0 | 2420 | 2959 | 3681 | 5195 | 5993 | 5711 | 8469 | 10003 | 9830 | 6935 | 4170 | 3924 | 4017 | 3951 | 3778 | 3440 |
| 60.0 | 2848 | 3036 | 4523 | 4920 | 6893 | 5178 | 7544 | 9236 | 9173 | 7503 | 4625 | 4372 | 4474 | 4460 | 4367 | 3593 |
| 62.5 | 3126 | 3279 | 4504 | 4170 | 4435 | 5227 | 7191 | 9742 | 9747 | 8576 | 4672 | 4236 | 4263 | 4029 | 3695 | 3243 |
| 65.0 | 3106 | 3228 | 4192 | 3966 | 4951 | 6174 | 7831 | 10818 | 10850 | 9316 | 3705 | 3089 | 3170 | 3128 | 2948 | 2790 |
| 67.5 | 2608 | 2829 | 3966 | 4287 | 5942 | 7045 | 8068 | 11441 | 11478 | 9436 | 3398 | 2741 | 2836 | 2824 | 2800 | 2649 |
| 70.0 | 2176 | 2339 | 3647 | 4501 | 5825 | 5816 | 7464 | 12308 | 12320 | 9226 | 3172 | 2510 | 2605 | 2615 | 2641 | 2471 |
| 71.0 | 1967 | 2130 | 3447 | 4299 | 5185 | 5122 | 7184 | 12401 | 12405 | 8907 | 3043 | 2408 | 2505 | 2522 | 2541 | 2352 |
| 72.5 | 1670 | 1831 | 3101 | 3510 | 3949 | 4104 | 6402 | 11887 | 11858 | 7948 | 2744 | 2174 | 2249 | 2215 | 2123 | 1821 |
| 75.0 | 1222 | 1341 | 2303 | 1938 | 2279 | 2629 | 4528 | 8204 | 8046 | 4170 | 808 | 443 | 506 | 572 | 701 | 626 |
| 77.5 | 735 | 784 | 1188 | 906 | 1229 | 1132 | 1349 | 3342 | 3333 | 1794 | 438 | 302 | 341 | 394 | 475 | 416 |
| 80.0 | 78 | 85 | 146 | 131 | 136 | 163 | 319 | 889 | 901 | 538 | 202 | 178 | 200 | 229 | 263 | 224 |
| 82.5 | 19 | 19 | 34 | 34 | 29 | 34 | 54 | 129 | 134 | 107 | 78 | 80 | 85 | 100 | 110 | 93 |
| 85.0 | 0 | 0 | 2 | 2 | 0 | 2 | 7 | 15 | 17 | 19 | 19 | 17 | 15 | 17 | 19 | 19 |
| 87.5 | 0 | 0 | 0 | 0 | 0 | 0 | 0 | 2 | 2 | 2 | 0 | 0 | 0 | 0 | 0 | 0 |
| 90.0 | 0 | 0 | 0 | 0 | 0 | 0 | 0 | 0 | 0 | 0 | 0 | 0 | 0 | 0 | 0 | 0 |

| Vertical Angles | Lateral Angles | | | | | |
|---|---|---|---|---|---|---|
| | 135.0 | 145.0 | 155.0 | 165.0 | 175.0 | 180.0 |
| .0 | 1644 | 1644 | 1644 | 1644 | 1644 | 1644 |
| 5.0 | 1872 | 1775 | 1687 | 1612 | 1570 | 1563 |
| 15.0 | 2529 | 2410 | 2181 | 1845 | 1546 | 1509 |
| 25.0 | 2661 | 2568 | 2434 | 2169 | 1624 | 1533 |
| 35.0 | 2678 | 2561 | 2434 | 2244 | 1675 | 1602 |
| 45.0 | 2734 | 2507 | 2342 | 2184 | 1702 | 1641 |
| 55.0 | 2873 | 2325 | 2169 | 2052 | 1724 | 1685 |
| 60.0 | 2680 | 2270 | 2053 | 1911 | 1612 | 1573 |
| 62.5 | 2503 | 2167 | 1999 | 1821 | 1512 | 1475 |
| 65.0 | 2371 | 2086 | 1911 | 1738 | 1407 | 1363 |
| 67.5 | 2247 | 1952 | 1665 | 1424 | 1166 | 1134 |
| 70.0 | 2021 | 1585 | 1176 | 964 | 854 | 842 |
| 71.0 | 1831 | 1375 | 983 | 779 | 723 | 716 |
| 72.5 | 1349 | 1005 | 679 | 596 | 577 | 575 |
| 75.0 | 458 | 484 | 514 | 526 | 516 | 516 |
| 77.5 | 299 | 331 | 365 | 404 | 414 | 414 |
| 80.0 | 175 | 197 | 219 | 260 | 278 | 278 |
| 82.5 | 83 | 88 | 95 | 124 | 127 | 127 |
| 85.0 | 15 | 15 | 12 | 19 | 17 | 15 |
| 87.5 | 0 | 0 | 0 | 0 | 0 | 0 |
| 90.0 | 0 | 0 | 0 | 0 | 0 | 0 |

**Illustration for lumen-point. Fig 3-14**

# Chapter Three  Lighting Engineering  91

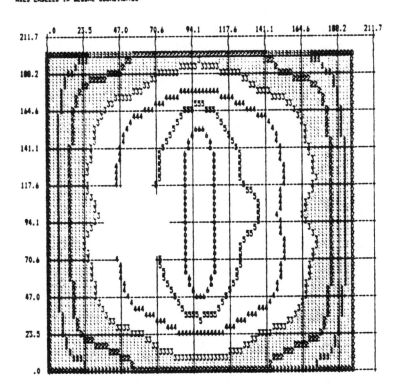

**Illustration for lumen-point. Fig 3-15**

## LUMEN-MICRO

LUMEN-MICRO performs the following functions: calculates at a rectangular grid of up to 20 x 20 points: horizontal and vertical footcandles, equivalent sphere illumination (ESI), room surface brightnesses, visual comfort probability (VCP) and relative visual performance (RVP). All the paramaters are user definable including size of grid and photometric data. It analyzes a mixed luminaire system, including direct, indirect, or direct/indirect equipment. With multiple overlays, the system can develop modes of irregular layouts of multiple luminaires. LUMEN-MICRO can account for the effect of areas of different reflectances on any wall surface. Models produced include point, line or area sources without assumption about the luminaire geometry. The user controls all aspects of the output report and can write-to-disk, plot, or print a custom report containing only the elements relevant to the current project. Finally, LUMEN-MICRO plots iso-contour or gray-scale renderings on an ordinary Epson or compatible graphics printer. A video module is available for Lumen-Micro which creates a color or grey-scale image directly on an AT&T Targa 16, Image Capture Board or an IBM EGA card.

### HOW THEY WORK

All the LUMEN family programs utilize worksheets into which the user enters data describing the environment, lighting array, luminaire photometric data and the definition of what analysis routines the user wishes performed. These sheets are then run individually if the floppy-disk based system is being used, or they can be run in batches if a harddisk is being utilized. This approach also allows modifying and rerunning at a later time. Each of the sheets is analyzed automatically by the system prior to running, and errors or omissions are noted for correction.

The analysis output is sent to an EPSON printer and is presented in text/columnar format, as well as with appropriate graphic presentations such as contour plots of equivalent sphere illumination or room surface representations. The output can also be analyzed with the data-conversion utility for Lotus 1-2-3.

Chapter Three   Lighting Engineering   93

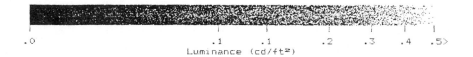

Modified view using corner downlights

Room Dimensions:

    X: 20.00
    Y: 20.00
    Z: 8.80

Observer Location:

    X: 11.30
    Y: 8.70
    Z: 4.50

OBSERVER

Viewing Direction:

    Horizontal Angle: -45.00
    Vertical Angle: 5.00

PICTURE PLANE PARAMETER

    Distance from Observer: 4.00
    Width: <6.10>
    Height: <4.00>

SHADING PARAMETERS

    Luminence at Lightest Shade: .50
    Percent White at Lightest: 100.00
    Percent White at Darkest: .00
    Print Density: quad

CALCULATED LUMINANCE AT ROOM SURFACES (cd/sqft):

|  | Maximum | Minimum | Average |
|---|---|---|---|
| North Wall: | .96 | .04 | .29 |
| East Wall: | .08 | .04 | .06 |
| South Wall: | .07 | .04 | .06 |
| West Wall: | .81 | .04 | .20 |
| Floor: | .39 | .00 | .08 |
| Ceiling: | .28 | .02 | .10 |

**Illustration of worksheet. Fig 3-16**

## PERSPECTIVE VIEWS

I would like to paraphrase from "Lumenews," published by Lighting Technologies, a description of a new computerized lighting tool they have developed. The system is based on the recursive equation and the "R-functions" of Iowa State University mathematician B.C.Carlson. This technique accounts for the bidirectional reflectance characteristic of the room surfaces. This is an alternative to ray tracing for the calculation of the luminances of room surfaces. This system operates on the first bounce. Second and subsequent bounce components are calculated using Fourier transforms. The results are combined to provide total luminance.

**Black and white perspective. Fig 3-17**

After developing the mathematical foundation of the problem, the user is allowed to describe the observer location and orientation, location and orientation of all luminaires, location and orientation of all surface, and luminance distribution of all surfaces. The end product of these techniques within the new microcomputer-based LumenMicro is a realistic perspective view of the lighting conditions in a room, in black and white or color!

**Perspective from targa. Fig 3-18**

This exciting design tool can evoke perception useful in the design process. According to Lighting Technologies, overall perceptions of a lighted environment depend upon the size, location, color, and luminance of all the objects in the visual field. Aside from scale, one can assume that if these pictures are a stimulus which has the same collection of sizes, location, and luminances as the real environment, the perceptions will be the same. This process doesn't depend on understanding how the perceptual process works, only that it is replicated in the model created by the computer. For example, the picture produced substitutes brightness for luminance. This is because brightness is a sensory response to the photometric stimulus luminance. The appearance of the resulting pictures,however, are dependent on the experience of the lighting designer for application and interpretation.

The resulting tool is a powerful asset to the development of viable humanistic lighting design solutions.

I have had an opportunity to work with a copy of the enhanced graphics module for LUMEN-MICRO and it is relatively easy to use. It requires only a minimal familiarity with the concepts of perspective drawing, just enough to define observer point and so forth. The resulting plots to the Epson printer accurately emulate the actual view of a lighted space. Plots of actual installations, when compared with black and white photographs of the identical plotted view, are astoundingly accurate! The ability to make multiple models of a lighting design in this way can serve as an important visualizing tool for the engineer, designer and scientist.

These products are being used by lighting engineers and designers, lighting manufacturers and suppliers, as well as by architects who are committed to the lighting design process. They are also used by lighting educators who find they are inexpensive and effective teaching tools. There are many lighting manufacturers who provide photometric databases for LUMEN-MICRO and related products, including Gardco Lighting, Peerless, Staff Lighting, and Lightolier. This commitment to excellence and thoroughness in the design process can only improve the lot of the end user of the lighting environment we designers create.

## SUMMARY

The engineering techniques covered in this chapter are sound and timeless. However, the methods and tools utilized by the lighting designer and engineer are rapidly changing. When I first started college the slide rule was the only way to do complex calculations. Computer modeling was still found only in the realm of the science fiction novel. Today, computer-aided drafting and project management are powerful aids to the designer. Never let your talent be thwarted by technology. Keep up with the latest computer programs and use them to the fullest. Remember, however, that the designer's imagination is the heart of the design process, and the computer is merely an efficient tool.

# Chapter 4

## LIGHTING TECHNOLOGY

We have made clear the concept that the human being and the technology interact in a lighting design and installation. The media the designer has to work with are all derived from technology. And technology is constantly changing. What I have attempted in this chapter is a snapshot of current technology with an emphasis on those elements that are fairly stable.

It is important to realize that one of the major jobs of a lighting designer is to remain up-to-date with technology. New products and materials as well as new engineering techniques are both a part of the novelty of the designer's pallette and a responsibility to your client. The aesthetic or novel is not, in itself, enough. New technologies for energy conservation and higher quality lighting environments are major elements of new product introductions. Attend trade shows, read magazines and devour product catalogs. It's a great part of the fun.

## LIGHT SOURCES

Artificial light may be produced in many different ways including: Incandescence, Photoluminescence and Electroluminescence. The first two are the primary producers of architectural lighting and the ones we will focus on.

### INCANDESCENCE

Incandescence is the emission of visible radiation produced by a body at a high temperature. Incandescent filament sources have a continuous spectrum as shown in the chart.

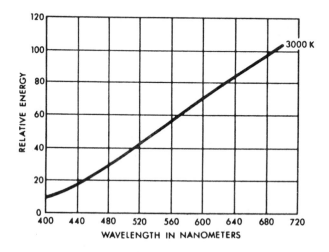

**Continuous spectrum. Spectral energy distribution in the visible region from a tungsten filament at 3000K.**
**Fig 4-1**

### PHOTOLUMINESCENCE

Photoluminescence results from the excitation of neutral gas atoms through collisions with electrons in an arc discharge. The resulting radiation may be visible to the eye (light), or invisible (ultraviolet). Radiation may be produced in a low pressure discharge (low pressure sodium lamp), or a high pressure discharge (high-intensity discharge [HID] light source family). In both cases the sources have line or broadened line spectra as shown in the chart.

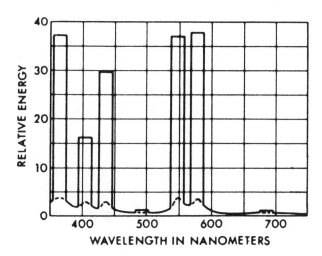

**Line Spectrum. Spectal Energy distribution typical of a clear mercury lamp**

**Fig 4-2**

## SOURCE CHARACTERISTICS

Light Output. One watt of radiant flux emitted at 550 nanometers (the point at which the eye is most sensitive) produces 680 lumens.

One watt of radiant flux distributed equally over the entire visible spectrum produces approximately 220 lumens--this is the theoretical maximum efficacy of white light.

Various classes of light sources approach these theoretical maxima in the manner illustrated in the chart.

Lamp Efficacy. The efficacy of a light source is the total luminous flux(lumens) emitted by that source divided by the power(watts) input to that source. It is expressed in lumens per watt.

Light Output Depreciation. All light sources depreciate as they age, i.e., the amount of light(lumens) they emit at any point along their burning life will be lower than emitted initially, This depreciation will vary with the source.

Mortality. All light sources fail in a predictable manner based on exhaustive tests. They vary depending upon the source.

Average rated life is the point at which 50 percent of the lamps in a given installation will have failed.

Mortality curves indicate the rate at which lamps are expected to fail for a large group.

Life. Source life will vary widely depending upon the characteristics of each.

Color. The color of a light source is dependent upon the total visible radiation emitted by that source. Source color is known a chromaticity.

|  | INCANDESCENT | | | FLOUR. | HIGH-INTENSITY DISCHARGE | | |
|---|---|---|---|---|---|---|---|
|  | Incan. | Tungsten-Halogen | Low Voltage |  | Mercury Vapor | Metal Halide | High-Press Sodium |
| SIZE & GEOMETRY | POINT | SMALL POINT | SMALL POINT | LINEAR | LARGE POINT | LARGE POINT | LARGE POINT |
| EFFICIENCY, LUMENS/WATT ("EFFICACY") | LOW 9-29 | LOW 20 | LOW 27-29 | HIGH 55-75 | MED 50 | MED 70 | HIGH 100 |
| LIFE, HOURS | SHORT 750-2000 | SHORT 2-4000 | SHORT 300-2000 | LONG 20,000 | LONG 24,000 | LONG 15,000 | LONG 24,000 |
| COLOR RENDERING | GOOD (WARM) AMBER | GOOD (CRISP) | GOOD | FAIR-GOOD 200+ COLORS | POOR | GOOD (COOL) BL-WH | POOR-FAIR (WARM) |
| INITIAL COST | LOW | MED. | MED. | LOW | HIGH | HIGH | HIGH |
| LIFE-CYCLE COST | HIGH | HIGH | MED. | LOW | MED. | MED. | MED. |

**Lamp Selection Chart. Fig 4-3**

## INCANDESCENT LAMPS

An incandescent lamp is one which emits light by heating a wire filament. Besides the wire filament it consists of a Bulb, or glass envelope, to protect the filament from oxidization, and a Base, used to connect the lamp to electricity and hold it in place. The incandescent lamp comes in a variety of shapes and sizes.

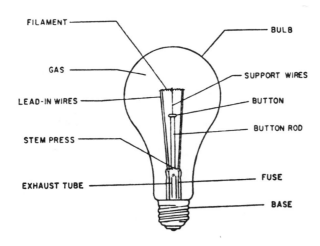

The incandescent filament lamp. Fig 4-4

## NOMENCLATURE.

Lamp designations include wattage, a letter which indicated bulb shape and diameter expressed in 1/8" increments, which is indicated by a number following the bulb shape designator. An example would be 200A23 which translates into a 200 watt "A" shaped lamp, which is the standard pear shaped "light bulb" used in the home, that is 2-3/8" in diameter. Technical literature on lamps also include the light center length (LCL) which measures the distance from the center of the filament to the tip of the base and the maximum overall length or (MOL). LCL is useful when the lamp is being used with reflectors or lenses and the placement of the filament affects the system.

102  Architectural Lighting Design

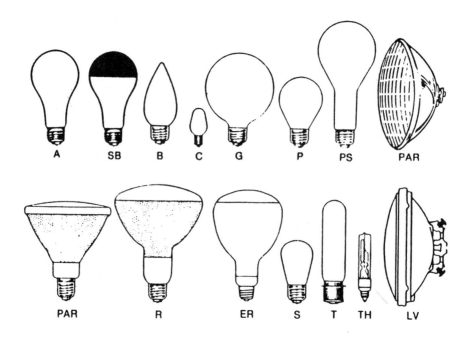

Bulb shape chart. Fig 4-5

## BASES

The base positions the filament and provides a mounting method for the lamp. It also makes the electrical connection. The most common base type is the medium screw base used on most home incandescent lamps. Others include Candelabra base, Mogul base Bipen and prefocus. See the chart below for shapes and sizes.

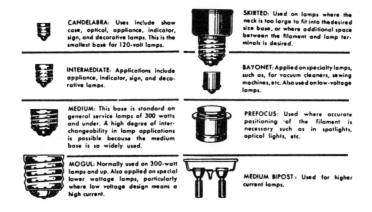

Base shape chart. Fig 4-6

## BULB FINISHES

Common finishes include clear, frosted, etched and coated. These finishes are added to help control the quality and distribution of the light. Colors are also often added to soften or change the light quality for special effects.

## EFFICACY

A typical rating for standard incandescent lamps is between 10 and 22 lumens per watt. Usually the higher wattage lamps are more efficient than the lower ones. For example a 25 watt lamp might produce 230 lumens while a 150 watt lamp might produce 2800 lumens; the efficiency of the 25 watt lamp is x lumens per watt while the 150 watt lamp is x lumens per watt.

## LAMP DEPRECIATION

This occurs by tungsten evaporation which causes bulb blackening. This problem is reduced in lamps larger than forty watts by filling them with an inert gas, usually a mixture of nitrogen and argon.

## COLOR TEMPERATURE

It is often important to know the color temperature in architectural lighting. The color temperature affects the spectra of the emitted light and the efficacy of the lamp. A chart representing the relationship between color temperature and efficacy is shown here.

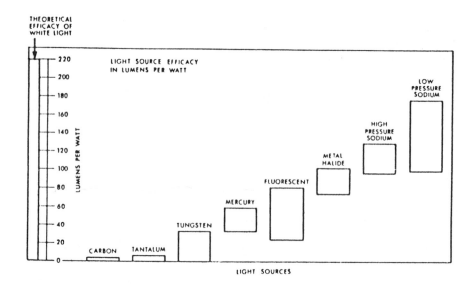

**Efficacy ranges of common light sources. Fig 4-7**

## CLASSES OF LAMPS

### GENERAL SERVICE

The lamps provide general lighting and come in sizes from 15 to 1500 watts.

### REFLECTORIZED LAMPS

Projector lamps. This class of lamps include the PAR or Parabolic Aluminized Reflector lamps which include a built-in lens system and a precisely positioned filament. This system makes an efficient beam control.

Reflector lamps or R lamps. Thes lamps are similar to PAR lamps but have a simple one piece bulb and do not have the precise beam control.

Silvered-bowl lamps. These lamps are generally available in A shapes and are used to reflect emitted light upward and reduce glare.

Semi-silvered-bowl lamps. These are similar to silvered-bowl lamps except there is a larger down component of light distribution.

Sign and Decorative Bulbs. These lamps are available in a variety of shapes, sizes and colors for special application.

Tungsten-Halogen Lamps. This is an important type of lamp in architectural design. They differ from conventional incandescent lamps in that they use the halogen cycle where evaporated tungsten returns to the filament, eliminating bulb blackening. The bulb is a high-silica tube which can withstand the high halogen cycle temperatures. See the chart below.

Special Application Lamps. This class of lamp includes Appliance Lamps, Rough Service Lamps, Vibration Lamps, Extended Service Lamps, Low Voltage Lamps and others.

## ADVANTAGES OF INCANDESCENT SOURCES

Concentrated filament is close to a point source which aids optical control.

- The circuitry is simple and requires no ballasts or electronic filters.

- Low initial cost.

- Rugged and operate in a wide variety of ambient temperatures.

## DISADVANTAGES OF INCANDESCENT SOURCES.

- Low efficacy.

    Give off a great deal of infrared which limits ultimate output. This also affects location and luminaire design.

- Critically affected by voltage variation.

- Short life.

## GAS DISCHARGE LAMPS.

This class of lamp includes fluorescent lamps, high-intensity discharge lamps (HID) and low pressure sodium lamps. The system produces light by the passage of an electric current through a gas like mercury vapor and a mixture of mercury and other metalic halides such as scandium iodide, sodium iodide, dysprosium iodide and sodium vapors. They produce light primarily invisible to the eye but which is used to excite a phosphor coating which then produces visible radiation. Low pressure sodium produces light directly by the excitation of the gas. These lamps produce line spectra rather than continuous spectra like incandescent lamps. This can greatly affect the selection of the efficient lamps in architectural placements. Excellent color rendition can be obtained by careful selection. See chart below.

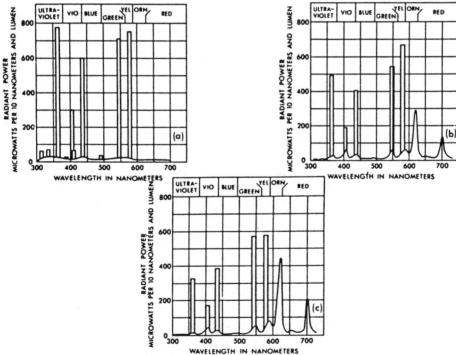

Spectral power distribution of light from typical mercury lambs (a) Clear mercury (b) delux white mercury. and (c) warm deluxe white mercury.

Figure 4-8

## FLUORESCENT LAMPS

A fluorescent lamp consists of a glass tube coated internally with a fluorescent material called a phosphor. This lamp is filled with argon or another inert gas and a small amount of mercury. The mercury vapor arc produces ultraviolet light which is changed into visible light when it collides with the phosphors. See the chart below.

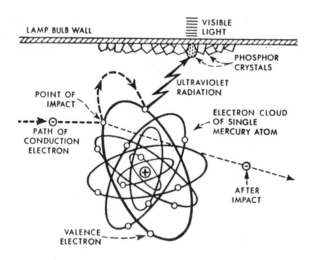

How light is produced in the fluorescent lamp: (a) Electrons emitted by electrode at one end of fluorescent lamp travel at high speed through the tube until they collide with one of the electrons of the mercury atom. (b) The impact diverts the electron of the mercury atom out of its orbit. When it snaps back into place, ultraviolet radiations are produced. (c) When the ultraviolet radiation reaches the phosphor crystals, the impulse travels to one of the active centers in the crystal and here an action similar to that described in Step (b) takes place. This time, however, visible light is produced.

Figure 4-9

The fluorescent TUBE can be straight, ranging in length from 6" to 96", or formed into a 'U' shape with legs 3" or 6" apart or made circular (circline fluorescent). The wattage of a fluorescent lamp usually indicates its nominal length which includes the lamp and two lamp holders. For instance, 20W tubes are 24" long, 30W are 36" and 40W are 48" long.

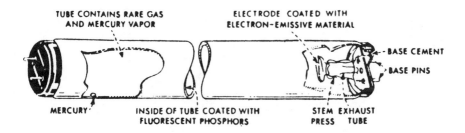

**Cutaway view of fluorescent lamp. Fig 4-10**

PHOSPHORS are chemicals that appear as the white powder on the inside of the tube. When activated by ultraviolet radiation, the phosphors give off light or fluoresce. Different colors and "shades of white" are produced by combining different phosphors in varying proportions. SED curves for some typical fluorescent lamp colors are:

110   Architectural Lighting Design

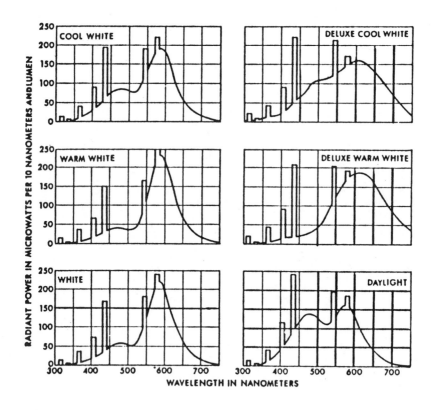

**Spectral distribution curves show lamp peaks at different points.
Figure 4-10**

ELECTRODES. There are two electrode types in fluorescent lamps; they are cold cathode and hot cathode. The hot cathode is a coiled tungsten filament which boils off elections at 1000 degrees centigrade to produce the arc. The cold cathode is made of iron or nickel. The cathodes use higher voltages. The standard architectural lamps are hot cathode. Cold cathode lamps are used in custom and special applications.

## LAMP DESIGNATIONS

Like incandescent, the nominal lamp Watts (exclusive of ballast watts), Shape and Diameter are generally indicated in the lamp designation preceded by 'F' for fluorescent and followed by designations of type and color. For instance, F40T12/RS/WW identifies a lamp as FLUORESCENT, using 40 WATTS, of a tubular SHAPE, 1 1/2" in DIAMETER, to be used with a RAPID START ballast and the color is WARM WHITE. There are some exceptions to this system such as the designation of the outside diameter of Circline lamps such as FC8T9 for a 22 watt, 81/4" diameter circular tube.

## TYPES OF FLUORESCENT LAMPS

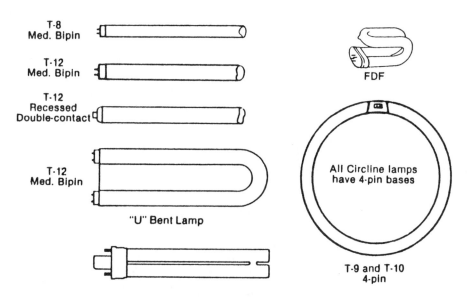

The fluorescent lamp is available in straight, U-shaped, and circular configurations and in various diameters.

**Fig 4-11**

AUXILIARY EQUIPMENT

Ballasts are required for fluorescent lamp operation to stabilize the lamp. Ballasts must match lamp type, wattage and line voltage to function properly. The average life of a ballast is about 15 years. For accuracy in calculating efficacy, ballast watts should be added to lamp watts.

PERFORMANCE CHARACTERISTICS

EFFICACY--Fluorescent lamps produce three to five times as much light as incandescent, ranging from 20 to 80 lumens per watt. Because they are so much more efficient, they produce much less heat. Generally, the longer the lamp, the greater its efficacy. Nominal lamp watts do not include power used by the ballast.

DEPRECIATION--Since fluorescent lamps depreciate rapidly during the first 100 hours of operation, definitive evaluation of footcandle levels should be taken only after the lamps are "seasoned", that is burned for at least 100 hours. Published lumen values are given after 100 hours of operation.

AVERAGE RATED LIVE--At three or more hours per start, ranges are from 7500 hours for a 15 watt lamp to 20,000+ hours for a 40 watt lamp. The fewer the starts, the longer the life.

EFFECT OF TEMPERATURE VARIATION--Fluorescent lamps are sensitive to ambient temperature. Light output changes when the wall of a tube is above or below optimum operating temperature (100°F). For example, heat build-up in a luminaire can reduce light output.

Special low temperature ballasts are required to start fluorescent lamps where the air temperature is below 50°F.

COLOR--A wide range of colors and "shades of white" is available because of different phosphor mixes. Color improved lamps, like the deluxe colors, have an added red component and are generally less efficient than standard colors. See lamp manufacturers' catalogs for proprietary and generic listings.

PREHEAT lamps employ a starting circuit to heat the cathodes before the arc strikes. Both a ballast and a starter are required. There is a delay of several seconds between the time the lamp is switched on and when it comes to full brightness. Preheat lamps are generally just the lower wattage fluorescents from 4 to 30 watts.

INSTANT START lamps use the ballast to supply high starting voltage thereby eliminating the need for a starter and speeding up the process. Most Instant Start lamps are the Slimline type which can be recognized by having a single pin rather than two pins projecting from the bases at each end.

RAPID START lamps, the most widely used type, combine properties of the preheat and instant start circuits. No starter is needed and the lamp achieves full brightness in 2 or 3 seconds without high starting voltage.

Standard Rapid Start lamps operate at 430 milliamperes (MA) or at about 10 watts per running foot. A "high output" lamp operates at 800 MA or 15 watts per foot providing about 50% more light. Highly loaded "extra high output" lamps operate at 1500 MA or 25 watts per foot and provide more than twice the light of a standard lamp.

COLD CATHODE instant start lamps have a longer life and lower efficacy than standard (hot cathode) fluorescent lamps. Special installation requirements must be met for in-door use of cold cathode because of its high starting voltage.

## EVALUATION OF FLUORESCENT LAMPS

| ADVANTAGES | DISADVANTAGES |
|---|---|
| 1. High efficacy | 1. Requires ballast |
| 2. Long life | 2. Temperature sensitive |
| 3. A linear source | 3. Size of lamp relatively big for amount of light produced |
| 4. Variety of colors | 4. Higher initial cost |
| 5. Low surface luminance | 5. Not suitable for critical light control |
| 6. Burns "cool" | 6. Can cause radio interference |

## HIGH INTENSITY DISCHARGE LAMPS

High Intensity Discharge (HID) lamps consist of an inner arc tube that contains gas vapors and electrodes and an outer jacket or bulb made of heat resistant glass. The outer bulb protects the arc tube, absorbs ultraviolet radiation from the arc and maintains a nearly constant temperature inside the lamp for proper operation. The outer bulb may be clear or phosphor coated. Light is produced when a high intensity electric discharge takes place in the gas vapor (fluorescent lamps use a low intensity arc).

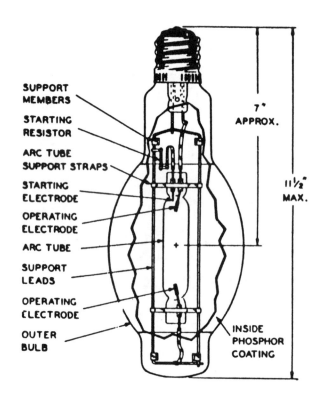

**A 400-watt phosphor-coated mercury lamp. Lamps at other sizes are constructed similarly.**
**Fig 4-12**

## TYPES OF HID LAMPS

MERCURY lamps produce a blue-green light suitable for some industrial and outdoor applications. Phosphor coated mercury lamps such as Deluxe White (DX) or Warm White Deluxe (WDX) not only produce vastly improved color qualities but may also add to the efficacy.

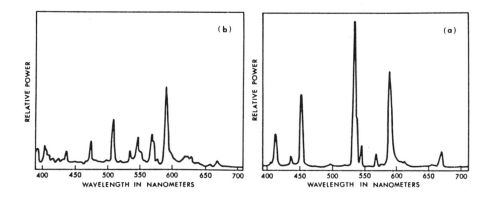

Spectral power distribution of (a) Sodium-inidium mercury iodide (b) Sodium-scandium mercury iodide.

**Figure 4-13**

SELF-BALLASTED MERCURY lamps have a ballast built into the lamp itself, eliminating the need for auxiliary equipment. Self-ballasted lamps have a shorter life and lower efficacy than standard mercury lamps.

METAL HALIDE lamps use various metallic halides as well as mercury in the arc, producing more light per watt and better color than mercury. Since the source of light, the arc, is relatively small, metal halide allows for good optical control. There is usually some color variation from lamp to lamp and over the period of life of a single lamp. Phosphor coating may be added to improve the color.

HIGH PRESSURE SODIUM lamps are the most efficient of the HID sources. They have a thin arc tube which allows for very good optical control. Sodium is the major chemical component hence the yellowish color of the light produced. It is generally not suitable where color rendering is important. However, a combination of HPS and metal halide lamps creates a balanced spectrum white light particularly suitable in malls and large scale atriums where plants are used extensively.

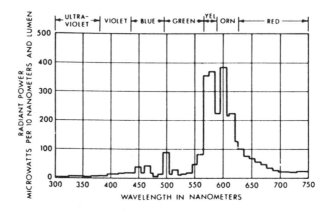

**The high-pressure sodium lamp spectrum is a broadened line spectrum with some of all colors present.**
**Fig 4-14**

## LAMP DESIGNATIONS

HID lamps have an American National Standard Institute (ANSI) designation: "H" for mercury, "M" for metal halide and "S" for high pressure sodium. In addition, lamp manufacturers may use proprietary names and designations. The information conveyed by a lamp code is as follows:

    H   Indicates lamp family
         (H-Mercury, M-Metal Halide, S-HP Sodium)

       33  Arbitrary numbers designating lamp
           electrical characteristics

          GL  Two arbitrary letters which describe
             lamp physical characteristics

             400-   Identifies lamp nominal wattage

                 DX-   Indicates phosphor color

    H33     GL-400/DX

Hence we know that a H33GL-400/DX is a 400 watt Delux White Mercury lamp, and a H38JA-100/WDX is a 100 watt Warm Deluxe Mercury.

## LAMP SHAPES

HID lamps are available in various bulb shapes:

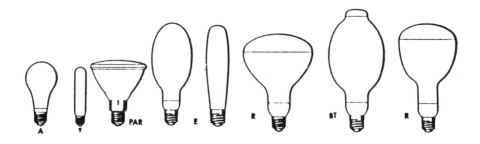

**Typical HID lamp bulb sizes. A-standard (Arbitrary), T-Tubular, PAR-Parabolic Aluminized Reflector, E-Elliptical, (Conical or Dimpled), R-Reflector, BT-Bulbous Tubular.**
**Fig 4-15**

## AUXILIARY EQUIPMENT

Like fluorescent, HID sources require a ballast for operation. The ballast must be matched to the lamp type, wattage and voltage of the lamp although some mercury and metal halide lamps may be used interchangeably on some ballasts. Metal halide ballasts will operate mercury lamps with no adverse effects. Lead-lag and series ballasts are also available in selected ranges from 5% to 15% of the lamp watts. To determine total wattage for an installation, add ballast watts to lamp watts.

## PERFORMANCE CHARACTERISTICS

EFFICACY--The most efficient HID sources are substantially more efficient than other commonly used sources, but there is a wide range of efficacy depending on the type and wattage. HID lamps range from 35 to 1000 watts and produce from 40 to 140 lumens per watt.

LIFE--HID sources have long lives thereby reducing maintenance costs. See lamp manufacturers' catalogs for specifics and Performance Characteristics Table for the range.

STARTING AND RESTRIKE TIME--HID lamps require some time to come to full brightness after being turned on. Further, if the lamp is turned off, it will not light (the arc will not strike) until it has cooled down (see Performance Characteristics Table). HID is not suitable where instant light is needed or expected as in a residential setting or where constant light is required for safety, as in a commercial interior. For this reason, HID installations are often supplemented by an incandescent emergency lighting system. HID lamps are sensitive to changes in voltage and may go out because of current variations.

| PERFORMANCE CHARACTERISTICS OF HID LAMPS | | | |
|---|---|---|---|
| | Mercury | Metal Halide | High Pressure Sodium |
| Wattages | 40 to 1000 W | 175 to 1500 W | 50 to 1000 W |
| Average Rated Life | 12,000 to 24,000 hrs. | 7500 to 20,000 hrs. | 12,000 to 20,000 hrs. |
| Efficacy | 30 to 65 LPW | 65 to 110 LPW | 100 to 140 LPW |
| Color--clear | poor | good | fair |
| Color w/phosphor | fair to good | good to excellent | ———— |
| Starting to full Brightness | 4 to 7 min. | 4 to 7 min. | 3 to 4 min. |
| Restrike Time | 4 to 7 min. | 4 to 7 min. | 1 to 2 min. |
| Cost | low | medium | high |

**Fig 4-16**

## BALLASTS

Discharge lamps require an auxiliary device known as a ballast to provide proper starting voltage and to limit the current and thus stabilize the lamp. A ballast is usually designed for a specific wattage and type of lamp and generally cannot be used with any other lamp. The exception is a mercury lamp which will operate satisfactorily on metal halide ballasts.

Essentially a ballast consists of an insulated wire (the COIL) wound around a metal CORE. A ballast may also have a capacitor to decrease the load on the wiring system by correcting the power factor (see below). Or it may have a RADIO INTERFERENCE-SUPRESSING CAPACITOR to reduce feedback that can cause static in nearby radios. A ballast generates heat. In an encased ballast, the space inside the case is filled with POTTING COMPOUND to improve heat dissipation and help reduce ballast noise.

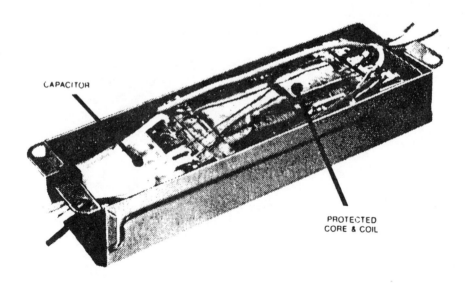

**Typical fluorescent ballast shown prior to filling with potting compound.**

**Fig 4-17**

## BALLAST WATTAGE

Ballasts consume power. Nominal lamp watts do not include watts used by the ballast. In calculating total electrical consumption of a fluorescent or HID installation, lamp and ballast watts must be added together. Some typical ballast wattages are:

| LAMP | BALLAST CONSUMPTION |
|---|---|
| 48" T-12, 40W, Preheat | 10W |
| 48" T-12, 40W, Rapid Start | 14W |
| High OUtput 40W flour. | 25W |
| Mercury - 100W | 18-35W |
| Mercury - 400W | 25-65W |

## BALLAST LIFE

Ballast life is determined by the ability of the insulation around the coiled wire to withstand the heaat generated. When ballasts are operated at their design temperatures, they should last about 15 years. Operation of ballasts at higher than design temperatures decreased ballast life. Generally, the cooler a ballast operates, the longer its life.

## POWER FACTOR

Current (amps) does not flow through the wire coil at the same rate of speed as through the metal core. This discrepancy is referred to as the power factor which represents the ratio of watts divided by volts times amps. Since volts and watts are usually fixed, amps (or current) will increase as the power factor decreases.

A low power factor ballast (40-50%) therefore requires more current hence greater wire size or fewer luminaires on a circuit. Adding a CAPACITOR helps equalize the flow of current resulting in a HIGH POWER FACTOR (HPF) of 90-95% and allows either the use of smaller circuit wires or more luminaires on a given circuit. High Power Factor ballasts are normally specified in commercial installations because of their lower current consumption. Low Power Factor ballasts are frequently used in residences because of their lower initial cost.

## BALLAST PROTECTION

If heat dissipation is not adequate, a ballast can become overheated, accelerating failure. Occasionally, but not very often, ballast failure may be accompanied by smoking, explosion or leaking. Fluorescent CLASS "P" ballasts contain a thermal protective device which deactivates the ballast when the case reaches a certain critical temperature. The device resets automatically when the case temperature drops to a suitable operating temperature. Class "P" ballasts are required for all interior fluorescent installations except for lamps which may be operated on simple reactance ballasts such as a preheated ballast. HID ballasts may be protected with either a fuse or a thermal protector similar to the Class "P" device.

## BALLAST NOISE

When alternating current flows through the winding around the core, it produces a low level hum. The noise level in an occupied space is dependent on the ballast mounting as well as the sound emitted by the ballast itself. Improper mounting can result in both amplification and transmission of the sound.

The noise levels of fluorescent ballasts have been rated on an alphabetic system ranging from 'A' (extremely quiet) to 'F' (quite audible). 'A' rated ballasts are commonly used in residences, schools and offices or wherever the hum would be considered distracting or annoying.

HID ballasts are rated by tests that determine SOUND PRESSURE LEVEL (lo). Normally Sound Pressure Level is represented as a NOISE CRITERIA (NC) number. NC sound ratings are shown as dashed lines on the Quick Calculator chart following.

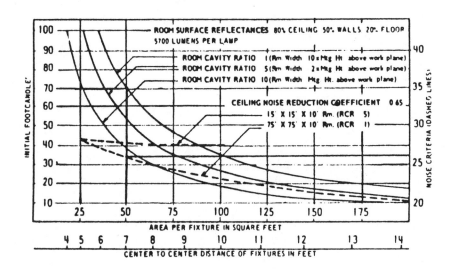

**HID Quick Calculator Chart. The sound rating varies with the room size and the placement of luminaires.**
**Fig 4-18**

## DIMMING

Light output of some fluorescent and some mercury lamps can be controlled by a DIMMING SYSTEM consisting of DIMMING BALLASTS and a DIMMING CONTROL. The dimming ballast replaces the standard ballast and must be matched to both the lamp and the dimming control. Factors that determine the type of dimming system best suited to a specific installation include: the number of lamps to be controlled, the dimming capability desired and the space required to install the controls and ballasts which are frequently larger than standard ballasts.

Systems for dimming metal halide and high pressure sodium lamp are not yet commercially available.

## FLUORESCENT BALLASTS

Fluorescent ballasts must be matched to the lamp wattage and type, i.e. preheat, instant start, rapid start, etc. Ballasts are available for both one and two lamp (lead-lag) operation. The latter eliminates any problems with stroboscopic (flickering) effect.

Removing one lamp from a luminaire with a two lamp ballast to save energy will result in improper operation of both the remaining lamp and the ballast as well as reducing ballast life. Removing both lamps from the circuit will save energy, but if the ballst remains energized, or wired into the circuit, it will continue to draw approximately 5 or 6 watts even without any lamps in the luminaire.

## ENERGY SAVING FLUORESCENT BALLASTS

There are three types of fluorescent ballasts available for reducing energy consumption. They are most commonly available for 40W lamps, standard or reduced wattage, for either 120V or 277V operation. They are:

1. Full Light Output-Reduced Wattage; the two lamp reduced wattage ballast cuts power consumption by 7 or 10 watts with about a 5% loss in light output.

2. Reduced Light Output-Reduced Wattage; one or two lamp ballasts provide 20 =-25% reduction in power consumption with a proportional decrease in light output.

3. Two Level Light Output-Two Level Wattage; these two lamp ballasts offer the option of operating lamps at full or at 50% output with a proportionate decrease in power consumption.

Energy Saving fluorescent lamps (Full Light Output-Reduced Wattage, 34W and 35W) used on Full Light Output-Reduced Wattage ballasts provide the optimum combination of watts consumed, light output and lumens per watt.

## HID BALLASTS

Ballasts for HID lamps may be incorporated into the body of a luminaire as a CORE AND COIL or ENCASED in its own metal compartment for use where separation of the ballast form the housing is desirable such as mounting the ballast in a pole or on a luminaire mounting frame.

## MERCURY BALLASTS

MERCURY BALLASTS- The most commonly used Mercury ballast is a CONSTANT WATTAGE AUTOTRANSFORMER ballast. Some advantages of this type over other types such as Reactor or

Autotransformer are: better regulation of lamp output, lower starting current and high power factor.

## METAL HALIDE BALLASTS

METAL HALIDE BALLASTS have a high power factor and will operated both metal halide and mercury lamps effectively. However, metal halide lamps will not operate reliably on most mercury ballasts. LEAD-PEAKED AUTOTRANSFORMER ballasts are the most commonly used for metal halide lamps.

## HIGH PRESSURE SODIUM

HIGH PRESSURE SODIUM lamps require a high voltage for starting. Both regulated and non-regulated HPF ballasts may be used with high pressure sodium lamps. Mercury and metal halide lamps cannot be used on HPS ballasts.

## SUMMARY

The technical apsects of the lighting installation and the technology available to the lighting designer are evolving at a rapid pace. Many of the techniques mentioned in this book will be out of date by the time it is published, and many new tools will be added to the designer's "bag of tricks." Keep up with new techniques and technologies. Add these new discoveries to your design notebook. Frequent lighting trade shows and read the magazines that support the lighting industry. Most important, talk to other designers, sales people, and technologists.

# Chapter 5

## LIGHTING GRAPHICS

### INTRODUCTION

This chapter is a collection of graphic symbols, forms, details, plans and construction documents. They are derived from my own practice and that of others. A great debt is due to Marlene Lee Lighting Design for a large number of the documents included here.

As I stated in chapter two, it is important to utilize the examples here and from other sources to learn from and adapt to your own situation. Borrow freely but be sure that details that you utilize are not mindlessly copied but rather are carefully modified to meet the needs on your current project. There is no such thing as appropriate "cookie cutter" design. Every design problem deserves your full attention and creativity.

On the other hand there is no such thing as isolated creativity. Every design owes its life to every other project the designer has created or experienced. This is as it should be. Human beings grow and change a small step at a time and aesthetic reality does the same.

Learn to be clear with your graphic and non-graphic design documents. Say everything you feel is important to your project. Those things you leave to the decisions of others may not come out the way you envision them. This may be satisfactory; some of my best designs have been influenced by contractors, electricians and owners. Don't be afraid to use whatever works. At the same time, if you do care, be specific and clear as to your intentions.

Use standard symbols for clear communication. There is a great deal of variety in graphic use and symbology but at the same time there is a limit. Just as a document with completely phonetic spelling may be clear it might also be confusing. Stay in the mainstream.

# SYMBOL LIST

|  |  |
|---|---|
| ⟶ | HOMERUN TO PANEL OR CABINET DESIGNATED |
| A-1,3,5 | THREE CIRCUITS TO PANEL "A" WITH COMMON NEUTRAL |
| ———— | CONDUIT RUN CONCEALED IN WALL OR ABOVE FINISHED CEILING - 1/2"C, 2#12 OR AS NOTED. |
| —‖‖‖— | 1/2"C, 3#12 (NUMBER OF CROSSMARKS INDICATE NUMBER OF #12 CONDUCTORS IN CODE SIZED CONDUIT. |
| — — — | CONDUIT IN OR UNDERFLOOR OR UNDERGROUND, WITH CONDUCTORS AS NOTED. |
| —×—×—×— | CONDUIT RUN EXPOSED |
| —T— | 3/4" TELEPHONE C.O. OR AS NOTED. |
| —T⊢— | 1" TEL. C.O. |
| —T⊢⊢— | 1 1/4" TEL. C.O. |
| Ⓐ flush | FLUSH MOUNTED LIGHTING PANEL ⎫ |
| surface | SURFACE MOUNTED LIGHTING PANEL ⎬ LETTERED BALLOON INDICATES PANEL DESIGNATION |
| power | SURFACE MOUNTED POWER PANEL ⎭ |
| | FLUSH & SURFACE MOUNTED CAB'TS AS NOTED. |
| ⊕ | DUPLEX CONVIENIENCE OUTLET MOUNTED +42" |
| ⊖ | DUPLEX CONVIENIENCE OUTLET, +12" OR AS NOTED. |
| ⊙ | DUPLEX CONV. OUTLET FLUSH IN FLOOR BOX |
| Ⓙ• | JUNCTION BOX, +12" OR AS NOTED |
| Ⓙ | JUNCTION BOX FLUSH IN CEILING OR ABOVE CEILING |
| ⊡ | FLUSH FLOOR BOX. SEE DRAWINGS FOR SIZE REQUIRED. |
| ▭〇▭ | RECESSED OR SURFACE MOUNTED FLUORESCENT FIXTURE WITH OUTLET BOX - SEE FIXTURE SCHEDULE FOR TYPE |
| ⊢〇⊣ | FLUORESCENT STRIP LIGHT |
| 〇 | RECESSED OR SURFACE MTD. FIXTURE WITH OUTLET BOX. |
| ()• | WALL MTD. FIXTURE WITH OUTLET BOX. |
| ↓⊗ | OUTLET & CEILING MTD. EXIT LIGHT ⎫ WITH DIRECTIONAL (SHADED PORTION INDICATES FACE) ⎬ ARROWS AS |
| ↓⊗• | OUTLET & WALL MTD. EXIT LIGHT ⎭ INDICATED |
| O | LIGHT OUTLET ON NITE LIGHT OR EMERGENCY CIRCUIT |
| $ | SINGLE POLE TOGGLE SWITCH +4° OR AS NOTED SUBSCRIPTS @ SWITCH INDICATE THE FOLLOWING: |
| | 2 - DOUBLE POLE          MC-MOMENTARY CONTACT |
| | 3 - THREE WAY            P - PILOT LIGHT |
| | K - KEY OPERATED         a,b,c - IDENTIFICATION OF OUTLET CONTROLLED. |
| [F]• | BREAKGLASS FIRE REPORTING STATION +5° |
| ▷[F]• | BREAKGLASS STATION WITH HORN ABOVE |

|  |  |
|---|---|
| ⊲ | TELEPHONE OUTLET, FLUSH IN FLOOR BOX. |
| ▷• | TELEPHONE OUTLET, +12" OR AS NOTED. |
| Ⓒ• | CLOCK & OUTLET +7º OR AS NOTED. |
| ▷• | INTERCOM OUTLET, +12" OR AS NOTED. |
| Ⓢ• | SPEAKER & OUTLET WALL MTD. +7º OR AS NOTED. |
| 600Ⓓ• | DIMMER-WALL MTD. +4º, RATING AS INDICATED. |
| Ⓣ• | THERMOSTAT OUTLET +5º |
| ⟨M⟩• | MICROPHONE OUTLET +12" OR AS NOTED. |
| ☒2 | MAGNETIC, LINE VOLTAGE STARTER (SIZE 2) |
| ☐☒ 70/3 | COMBINATION STARTER WITH BREAKER - 70A-3P |
| F☐ | EXO NON-FUSED DISC. SWITCH          (F-FUSED) |
| ☐ | CONTROL DEVICE AS INDICATED |
| Ⓢ | MOTOR OUTLET, HORSEPOWER AS DESIGNATED. |
| Ⓕ | FAN OUTLET |
| ○—— | CONDUIT UP |
| •—— | CONDUIT DOWN |
| ⊢—— | CONDUIT STUBBED OUT & CAPPED. |
| ▭▭▭ | PLUG-IN-STRIP WITH OUTLETS AS NOTED |
| Ⓐ | FIXTURE TYPE DESIGNATION. SEE FIXTURE SCHEDULE |
| ⊖• | SPECIAL RECEPTACLE. SEE DRAWINGS FOR TYPE REQ'D. |

## ABBREVIATIONS

| | |
|---|---|
| C.O. | CONDUIT ONLY |
| N.F. | NON FUSED |
| W.P. | WEATHER PROOF |
| M.H. | MOUNTING HEIGHT |
| E.P. | EXPLOSION PROOF |
| I.C. | INTERRUPTING CAPACITY |
| C | CONDUIT |

# Architectural Lighting Design

## SINGLE-LINE DIAGRAM SYMBOLS

**100 FB 3P** — THERMAL MAGNETIC MOLDED CASE CIRCUIT BREAKER WITH FRAME AND TRIP RATING AND NUMBER OF POLES INDICATED. (NA INDICATES NON-AUTOMATIC.)

**100L FB 3P** — THERMAL MAGNETIC MOLDED CASE CIRCUIT BREAKER WITH INTEGRAL CURRENT LIMITERS, (PROVIDE STANDARD CURRENT LIMITERS UNLESS OTHERWISE INDICATED), WITH FRAME AND OVERLOAD TRIP RATING AND NUMBER OF POLES INDICATED.

**SPCB 600 400** — MOLDED CASE CIRCUIT BREAKER WITH CURRENT MONITORS, STATIC SENSOR AND SHUNT TRIP, WITH FRAME AND CURRENT MONITOR RATING INDICATED. (SEE NOTE 1 FOR STATIC SENSOR REQUIREMENTS.)

**7 MCP 3P ⊠1** — COMBINATION MOTOR CIRCUIT PROTECTOR WITH FRAME AND TRIP RATING INDICATED AND MAGNETIC MOTOR STARTER WITH THREE OVERLOAD RELAYS NEMA SIZE AS INDICATED.

**GFS 1200A** — CURRENT TRANSFORMER AND GROUND FAULT SENSOR WITH PICK-UP RATING INDICATED. ASSOCIATED CIRCUIT BREAKER TO HAVE SHUNT TRIP.

**—Ⓐ—** — AMMETER WITH SELECTOR SWITCH AND CURRENT TRANSFORMERS AS INDICATED.

**—Ⓥ—** — VOLTMETER WITH SELECTOR SWITCH.

**2000A - 277/480V** — MAIN OR DISTRIBUTION SWITCHBOARD BUS CURRENT AND VOLTAGE RATING

**2½"C, 4#4/0** — INDICATES FEEDER CONDUIT AND CONDUIT SIZE

**(15)** — MOTOR - HORSEPOWER INDICATED.

**A** — PANELBOARD. 'A' INDICATES PANEL 'A'. SEE SCHEDULE OF PANEL FOR REQUIREMENTS.

## GENERAL NOTES

1. STATIC SENSORS SHALL HAVE THE FOLLOWING FEATURES: 1.) LONG TIME DELAY, 2.) SHORT TIME DELAY, 3.) INSTANTANEOUS TRIP, 4.) GROUND FAULT TRIP.

## FIXTURE SCHEDULE

| TYPE | DESCRIPTION | LAMPS | MANUFACTURER |
|------|-------------|-------|--------------|
|      |             |       |              |

## Chapter Five  Lighting Graphics  131

**PROJECT:**
**PROJECT NO:**
**DATE:**
**BY:**

**Marlene Lee**
LIGHTING DESIGN
CONSULTING ENGINEERS
SAN FRANCISCO  HOUSTON

573 MISSION STREET
SAN FRANCISCO, CA 94105
TELEPHONE (415) 546-7885

LUMINAIRE SCHEDULE
Page ____ of _____

| TYPE | DESCRIPTION | FIXTURE Manufacturer/Product | LAMP Ordering Code | INPUT WATTS* |
|------|-------------|------------------------------|--------------------|--------------|
|      |             |                              |                    |              |

\* includes ballast losses

Architectural Lighting Design

Marlene Lee

LIGHTING CALCULATIONS

| ROOM | | | | | | REFLECT | | | LUMINAIRE | | | | | ILLUM |
|---|---|---|---|---|---|---|---|---|---|---|---|---|---|---|
| NO/NAME | AREA | L | W | H | RCR | F | W | C | TYPE | CU | NO | LUMENS | LLF | |

# EXAMPLES

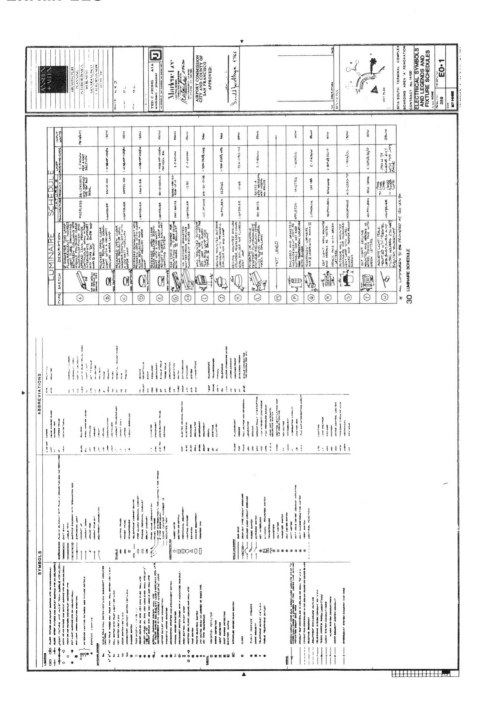

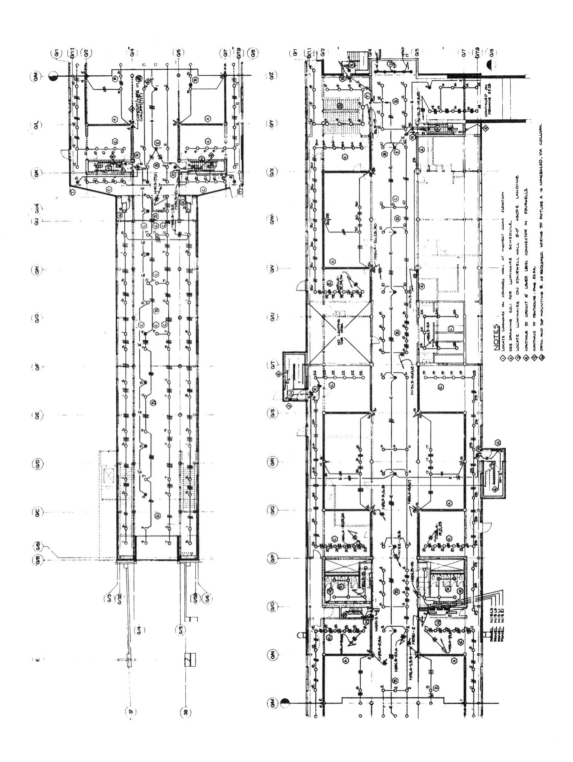

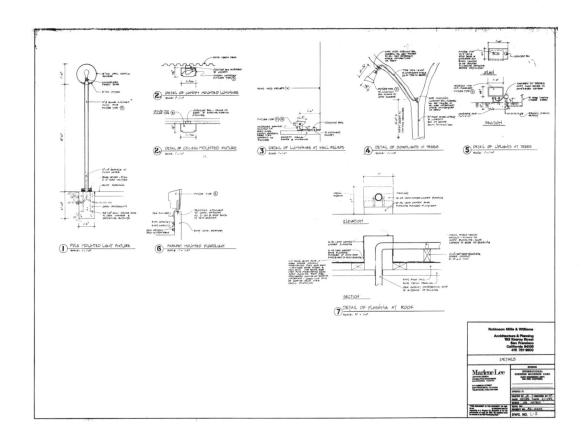

# Architectural Lighting Design

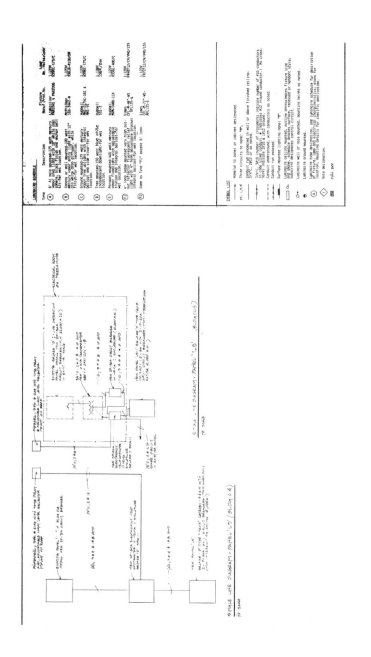

# Chapter Five  Lighting Graphics  137

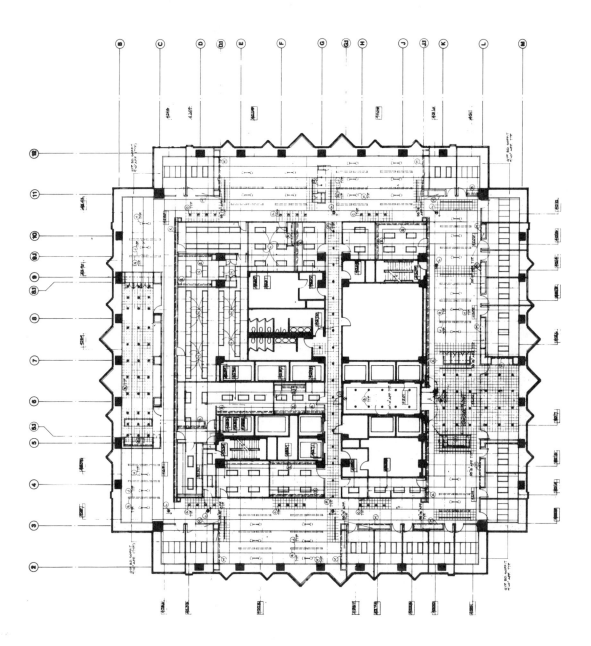

# 138 Architectural Lighting Design

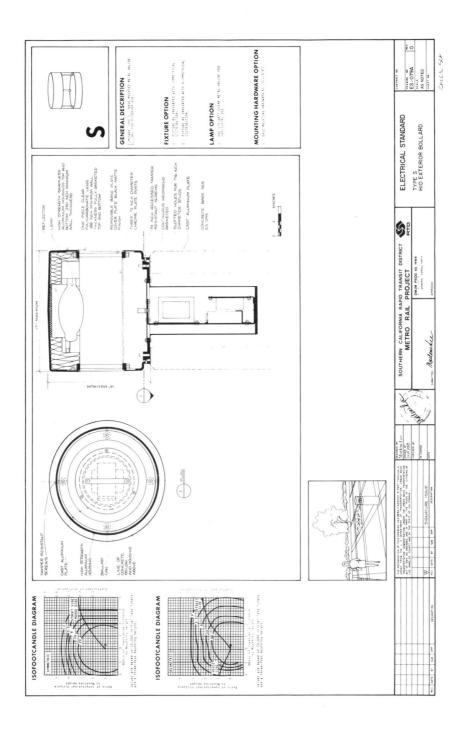

# Chapter Five  Lighting Graphics

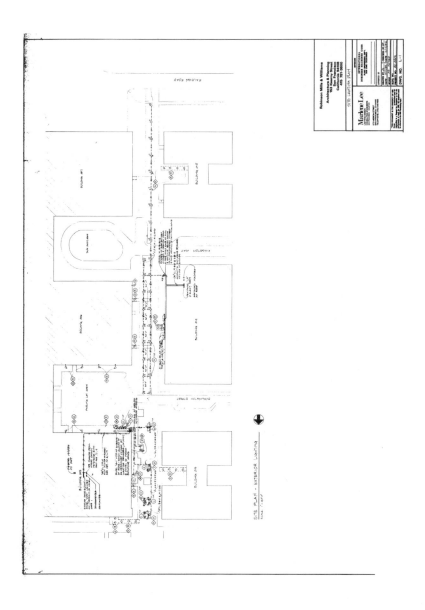

## Architectural Lighting Design

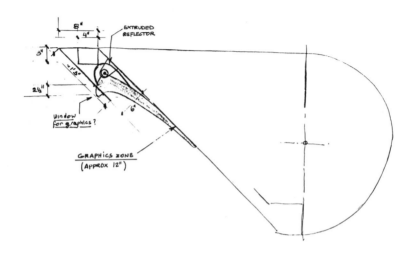

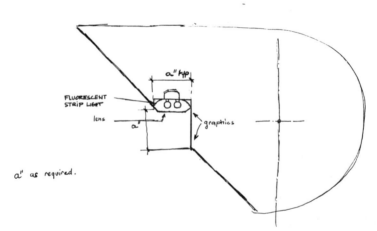

LIGHTING + GRAPHICS / AIR SUPPLY DEVICE
SCALE: 1½" = 1'-0"   **D**
8306 23/5

**Marlene Lee**
LIGHTING DESIGN
CONSULTING ENGINEERS
SAN FRANCISCO  HOUSTON

573 MISSION STREET
SAN FRANCISCO, CA 94105
TELEPHONE: (415) 546-7885

# Chapter Five  Lighting Graphics

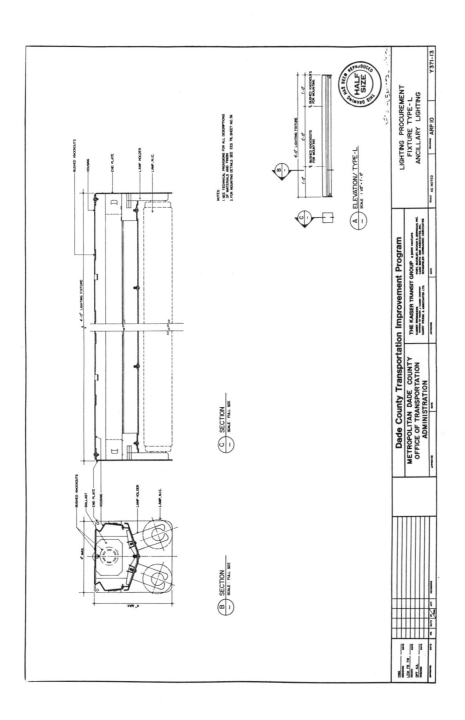

## 142  Architectural Lighting Design

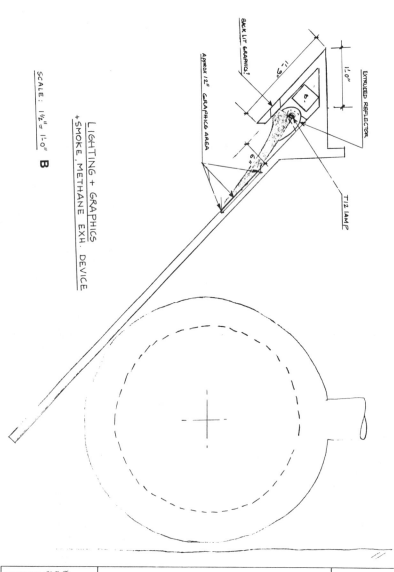

# Chapter Five  Lighting Graphics

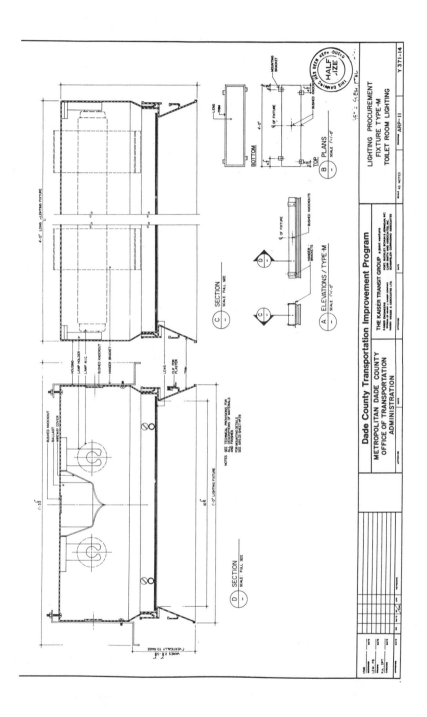

# 144  Architectural Lighting Design

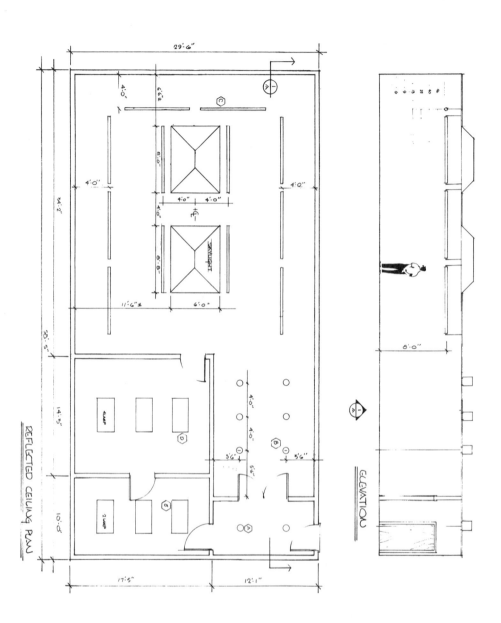

# Chapter 6

## SPECIFICATIONS

### INTRODUCTION

This is a sample specification developed by Marlene Lee Lighting Design and modified for each project depending on the equipment and conditions of both the building and the contract. It is included here as an example of specification writing for the lighting designer. It is important when writing specifications that the designer NOT simply copy a master specification but rather read and rewrite EVERY section that is different from or in conflict with the current project. The specification is part of the legal building contract and is just as important as the project manual, drawings and bills-of-material. The sample specification section is intended to be only a model and the author and his source in no way take responsibility for its use.

### SAMPLE SPECIFICATION

SECTION 16500

LIGHTING

PART 1 - GENERAL

*Changes and additions will be required to modify this Master Specification to reflect the specific requirements of a project. The specifier shall identify such modifications and supplemental requirements to this Section for inclusion in the project specification.*

*Coordination of Specifications and Drawings*

*This Master Specification was developed with the assumption that the drawings for a specific job will be coordinated with the specifications, and that the drawings will include the following:*

1. *Lighting Fixture Schedule with each luminaire type identified in the same way as in the specifications, e.g., Type "A", Type "B". The Lighting Fixture Schedule will contain requirements for fixture materials and finishes.*

2. Plan locations of luminaires.

3. Details for mounting luminaires to associated environmental systems, ceilings, and supporting structures.

4. Control diagrams.

*Lines such as this one are assistance guides to the specifier which should be deleted before publication.*

*An underlined blank space preceded by a bracketed capital letter -B- indicates the choice or requirement of an insertion necessary to the project. Cross out such indications where an insertion is not required for your project.*

1.01 DESCRIPTION

A. This Section specifies the furnishing and installation of lighting systems, complete and operable, as indicated, including fixtures; fixture mounting hardware, including brackets, canopies, hangers, and poles; lamps; auxiliary lighting equipment; and lighting control equipment.

B. The equipment and materials required under this Section shall be in accordance with the general description as indicated in Section: BASIC MATERIALS AND METHODS.

*Cited references should be modified to include only those references invoked after you have edited this section for the lighting requirements of a particular job and must include reference to a particular article, publication, etc. References italicized should be eliminated if specific sections are not cited.*

1.02 CITED REFERENCES

A. AA - Aluminum Association

B. ANSI - American National Standards Institute

01. C78; Incandescent Lamps/Electric Discharge Lamps (fluorescent)/ Electric Discharge Lamps (Mercury) / High-Intensity Discharge Lamps/Fluorescent Lamp Auxiliaries

02. C81; Electric Lamp Bases and Holders

03. C82; Lamp Ballasts

C. ASTM - American Society for Testing Materials

   01.  A 167; Spec. for stainless and heat-resisting Chromium-Nickel Steel Plate, Sheet and Strip

   02.  A 366; Spec. for Steel, Carbon, Cold-rolled Sheet commercial Quality

   03.  A 386; Specification for Zinc Coating (Hot Dip) Assembled Steel Products

   04.  A 584; Specification for Copper Alloy Sand Casting for General Application

D. CDA - Copper Development Association

E. FS - Federal Specification

   01.  W-S-896;

F. MIL - Military Specifications

G. NEMA - National Electrical Manufacturers Association

H. NFPA - National Fire Protection Association

   01.  70; National Electric Code

   02.  101; Life Safety Code

I. UL - Underwriters' Laboratories, Inc.

   01.  57; Electric Lighting Fixtures

   02.  496; Edison-Base Lampholders

   03.  508; Electric Industry Control Equipment

   04.  542; Lampholders, Starters and Starter Holders for Fluorescent Lamps

   05.  773; Plug-in Locking-Type Photocontrols for Use with Area and Roadway Lighting

   06.  887; Delayed-Action Timelocks

07. 924; Emergency Electric-Lighting Equipment

08. 935; Fluorescent-Lamp Ballasts

09. 1029; High-Intensity-Discharge-Lamp Ballasts

J. PEI - Porcelain Enamel Institute

01. S-100; Specification for Architectural Porcelain Enamel on Steel for Exterior Use

02. ALS-105; Recommended Specification for Architectural Porcelain Enamel on Aluminum for Exterior Use

1.03 TESTS AND ACCEPTANCE

A. This section specifies the documentation required and the factory tests to be conducted on the specified lighting fixtures.

B. The lighting fixtures to be tested shall be typical of the unit it represents, clean and free of mechanical defects, equipped with the proper fittings, and with the lamp of the size and type in the position recommended for service operation.

C. Material, equipment and components shall be tested in accordance with UL standards, and shall be UL listed. Material, equipment and components not covered by UL standards shall be tested in accordance with other nationally recognized standards approved by the Designer, or shall be of the kind whose production is periodically inspected by a nationally recognized testing laboratory approved by the Designer, and shall bear a label tag or certification of such inspection.

D. Tests for photometric performance shall be made and reported in accordance with the approved methods outlined by the IES for photometric testing, and shall include data on candle-power distribution, zonal lumens, and maximum luminance values and luminaire efficiency including complete coefficients of utilization tables to indicate compliance with performance requirements.

E. All test data shall be reported on 8-1/2 by 11 inch sheets and shall be certified by a nationally recognized independent testing laboratory.

1.04 QUALITY ASSURANCE

This section should be edited to reflect specific project requirements and budget.

A. This section specifies the submittals required to show compliance with the intended quality of design, fixture construction, installation and maintenance procedures.

B. A sample of each type lighting fixture representative of the Manufacturer's typical product and of material and finish indicated shall be submitted. Electrical components shall be as indicated with the exception of ballasted luminaires indicated to be operated at 277 volts which shall be supplied for test operation at 120 volts. All indicated fixture hardware and lamp(s) shall accompany the sample and the sample shall be provided for operation at 120 volts with a six-foot cord and plug.

   01. Each sample shall have its materials identified; i.e., alloy, thickness, gauge, temper.

   02. Variations in the finish colors of the samples submitted shall not exceed the color range established by the accepted samples on display at the Designer's Sample Room.

   03. Processing of production materials shall not begin until Designer's written approval of samples has been obtained.

C. A Luminaire Manual shall be submitted which provides all documentation indicating fixture construction, installation and maintenance.

   01. The Manual shall be complete with cover, title page, and table of contents. The cover and title page shall identify the document, project, client, contract name, number and date of issuance. The table of contents shall provide at a glance the overall document scope and structure and, as a minimum, a heading for each fixture type with subheadings for each drawing and test report. Documents shall be grouped by fixture type with each grouping prefaced by a "general information" report sheet. Blank report sheets will be provided by the Engineer.

   02. Documents and drawings submitted will be required to be of a quality which will facilitate fully legible reproduction of all information through a sequence of reproduction to 35 mm silver halide film processed to archival standards and further reproduced to original size by normal printing processors. All submittals for record or approval shall be full size reproducibles. Updated and/or revised drawings shall be equal in quality to

original submittals and processed in the same manner. The Contractor shall not change drawing numbers when redrawing for clarity or updating. A superseding drawing shall be labeled as redrawn with revision and date appropriate to the superseded drawing.

03. Luminaire construction shall be sufficiently detailed to show overall dimensions and material composition. It shall indicate housing, light controlling elements, finishes, electrical components including lampholders and ballasts and fixture hardware.

04. Procedures shall be clearly indicated for the installation of the complete lighting unit in its final service location. Templates shall be provided as indicated. Locations of all parts and openings interfacing with remote systems; i.e., poles, bases, mounting hardware, auxiliary electrical equipment, lighting control equipment, and lamps shall be dimensioned.

05. Maintenance procedures shall be described including: 1) all materials and components clearly indicated in a parts list, 2) relamping methods, 3) special tools required, and 4) frequency of inspection, tightening or other service recommended for preventative maintenance.

06. All test reports specified in Section 1.03 shall be included.

1.05 SPARE PARTS

A. This section specifies requirements for furnishing spare parts.

B. All spare parts shall be packaged in accordance with the General Provisions. Guaranteed availability of spare parts shall be 10 years after expiration of warranty.

C. All spare parts shall be furnished as specified in the Bid Form. A recommended spare parts list with unit price and effective pricing dates shall be submitted.

1.06 PACKAGING AND HANDLING

A. Products shall be handled and transported in a manner that prevents damage to the same.

B. Products shall be wrapped and packaged to avoid damage to the same. A packaging plan shall be submitted for review and approval.

C. Each carton shall be indelibly marked with minimum 1/2 inch high letters with the following information:

01.  Fixture, Lamp or Component Type

02.  Quantity

03.  Manufacturer's Name and Product Number

## PART 2 - PRODUCTS

### 2.01 LIGHTING FIXTURES

*The lighting fixture schedule lists three trade names to identify product. The use of a single trade name, followed by the words "or equal", is permissible only when no other trade name can be found. Identify each item with the name of the manufacturer, model number, type, size, finish and other applicable classifications. When the manufacturer is not nationally known, include his address or the name and address of a local distributor in the reference.*

A. General Requirements

01.  Lighting fixtures shall be provided, complete and ready for service, in compliance with UL 57, of the number, type, material, finish, electrical components and characteristics, and with all necessary hardware and auxiliary equipment indicated.

02.  The fixtures shall be clearly marked with manufacturer's name and catalog number, voltage, acceptable lamp type and maximum wattage, and labeled for the intended use.

03.  Fixtures, other than those for interior use only, shall be rain-tight and dust-tight.

B. Materials

*Where a separate Section of the Specifications contains standards and general specifications for Architectural Materials, i.e., Architectural Aluminum, this specification should make reference thereto.*

01.  Thicknesses, gauges and tempers of products shall be as indicated, and as recommended by the manufacturer for the

specific finish, proper forming operations and structural requirements.

02. Lighting sheet for reflector material: Prefinished aluminum, minimum thickness 0.032 inch, Architectural Type 1 with Class M1, anodic coating providing 83 percent reflectivity.

03. Concrete for base foundations shall be Section: Portland Cement Concrete; Section: Concrete Reinforcement; Section: Precast Concrete.

04. Stainless steel shall be type 316 conforming to ASTM A 167.

05. Acrylic for lenses and diffusers shall be manufactured from virgin-acrylic extrusion or injection molding pellets.

06. Fiberglass panels for diffusers shall be cast from pure methyl methacrylate monomer reinforced with fiberglass, minimum 1/8 inch thickness.

07. Polycarbonate for windows shall be manufactured from resin designed for use with HID lamps.

*Surface finishes depend on the alloy selected, mechanical processes, chemical processes and specified coatings.*

C. Finishes

01. Lighting fixtures shall be completely factory-finished in colors to match samples in the Designer's Sample Room and in accordance with the manufacturer's recommendations for the specific application.

02. Commence no finishing operations until fabrication and forming operations have been completed.

03. Aluminum work to be anodized shall be given as described by the Aluminum Association, a preanodic treatment followed by an Architectural Class 1, anodic coating.

    a. Aluminum shall be anodized in accordance with procedures established by alloy manufacturer to achieve color within specified range,

    b. A clear organic protective coating shall only be applied to exposed aluminum surfaces that may experience prolonged contact with caustic material, i.e., concrete, plaster.

04. Minimum cleaning of metal prior to painting shall be a 5-stage phosphatizing system as stated in the accepted submittals.

05. Interior fixtures with surfaces not exceeding 150-degrees F. shall be painted two coats, minimum, of either alkyd or acrylic gloss enamel to a minimum total dry film thickness (DFT) of 2.5 mils.

06. Interior fixtures with surfaces exceeding a temperature of 150-degrees F., but not exceeding 300-degrees F., shall be painted with silicone-alkyd enamel, two coats minimum to a total dry film thickness (DFT) of 2.5 mils, minimum.

07. Fixtures specified to be coated shall be provided one coat of epoxy-polyamide at a minimum dry film thickness (DFT) of 2.0 mils and one coat of aliphatic urethane to a minimum DFT of 2.0 mils. Interior reflective surfaces specified to be painted shall be as specified for interior fixtures.

08. Fixtures specified to be porcelain enamelled, or painted fixtures with reflectors specified to be porcelain enamelled shall be finished in accordance with the requirements of PEI-S-100, or PEI-ALS-105.

09. Reflective surfaces not specified to be specular shall be gloss white, guaranteed non-yellowing, with a reflectance rating of not less than 88 percent.

10. Galvanized coating: Hot-dip galvanized or hot-zinc coating according to ASTM A 386. Where painting of the galvanized surface is indicated, the surface shall be prepared with vinyl acid wash primer with polyvinyl butyral resin 56 pounds, 80 gallons zinc chromate pigment and phosphoric acid.

D. Electrical Components

  01. Lampholders

    a. General Requirements

      (1) Lampholders and sockets shall be of the class and style recommended by the lamp manufacturer for the specific lamp required by each fixture design and rated for 660 watts, 600 volts or as indicated.

(2) Lampholders and sockets shall be rigidly and securely fastened to the mounting surface with the necessary provisions to prevent lampholder from turning and shall be front removable without dismantling any part of the fixture.

(3) Lampholders and sockets shall be correctly located in the lighting fixtures to place each lamp, of size specified, in proper position with relation to the fixture design specified. They shall be clearly marked to indicate manufacturer lamp type and voltage and appropriate listings.

b. Incandescent and high intensity discharge (HID) lampholders shall have a glazed porcelain body with nonferrous metal components of heavy duty design, and be vibration resistant. Edison-based lampholders to be in accordance with the applicable requirements of UL 496.

(1) General purpose incandescent lamps use a medium screw base socket.

(2) Incandescent mogul base lamps of PAR configuration use a mogul end prong base socket rated 1000 watts, 125 volts.

(3) Special low wattage incandescent lamps such as the 20-watt T6-1/2 use a phenolic bodied dc bayonet socket rated 75 watts, 125 volts.

(4) Single ended tungsten-halogen lamps use a minican screw socket.

(5) Double ended tungsten-halogen lamps use recessed contact socket rated 3000 watts, 600 volts.

(6) Mercury lamps of the B, R or PAR configuration up to 175 watts use a medium screw base socket.

(7) Mercury or metal halide lamps that are to be operated in the horizontal position use a position oriented mogul base socket.

(8) High pressure sodium lamps up to and including 1000 watts use a high voltage mogul lampholder, 5Kv pulse rated, 1500 watts, 600 volts.

c.  Fluorescent lampholders shall be of white urea, spring loaded with silver-plated contacts of the pedestal or butt-on type, in accordance with the applicable requirements of UL 542.

(1) Rapid start (430Ma) lamps use medium bipin spring loaded lampholders of the tombstone or butt configuration.

(2) Rapid start (800 and 1500Ma) lamps use recessed double contact lampholders of the telescopic type.

(3) Instant start, slimline lamps, use single pin lampholder of the telescopic type.

02.  Ballasts

a.  Ballast operating characteristics shall comply with the recommendations of the lamp manufacturer with regard to lamp electrical characteristics. Ballasts shall be suitable for the line voltage with 0.9 power factor, and maximum current crest factor of 1.8. The ballast shall provide reliable lamp starting at the minimum temperature indicated. Operate in ambient temperatures up to 105 degrees F with maximum ballast case temperature of 90 degrees C. Each ballast shall be securely mounted inside the fixture, in such a manner as to obtain the necessary heat dissipation. High intensity discharge ballasts shall conform to the applicable requirements of UL 1029. Fluorescent ballasts shall conform to the applicable requirements of UL 935.

b.  Mercury lamps: Operated by a constant wattage auto-transformer, CWA, type ballast. The ballast shall provide reliable single lamp starting at minus 20 degrees F, and allow plus or minus five percent lamp watts variation for a plus or minus 10 percent input voltage variation.

c.  Metal halide lamps: Operated by a lead peaked auto, LPA, type ballast. The ballast shall provide reliable single lamp starting at minus 20 degrees F, and allow plus or minus 10 percent lamp watts variation for a plus or minus 10 percent input voltage variation.

d.  High pressure sodium lamps 150 watt size and smaller: Operated by a high leakage reactance, HX type ballast. The ballast shall provide reliable lamp starting at minus 20 degrees F, and allow plus or minus 12 percent lamp watts variation for a plus or minus five percent input voltage variation.

e. High pressure sodium lamps, 250 watt size and larger: Operated by a constant wattage autotransformer, CWA type ballast. The ballast shall provide reliable lamp starting at minus 20 degrees F, and allow plus or minus five percent lamp watts variation for a plus or minus ten percent lamp watts variation for a plus or minus ten percent input voltage variation.

f. Rapid start, 430 Ma, fluorescent lamps: Operated by a Class P, 430 Ma, 60 hertz ballast. The ballast shall provide reliable lamp starting at 0 degrees F, and shall be provided with a sound level rating of "A". Lamp voltage variation shall not exceed plus five percent and minus 10 percent.

g. Rapid start, 800 Ma, high output fluorescent lamps: Operated by a Class P, 800 Ma, 60 hertz ballast. The ballast shall provide reliable lamp starting at minus 20 degrees F, and shall be provided with sound level rating of "B" or better. Lamp voltage variation shall not exceed plus five percent and minus 10 percent.

h. Rapid start, 1500 Ma, fluorescent lamps: Operated by a Class P, 1500 Ma, 60 hertz ballast. The ballast shall provide reliable lamp starting at minus 20 degrees F, and be provided a sound level rating of D or better. Lamp voltage variation shall not exceed plus five percent and minus 10 percent.

i. Slimline and instant start lamps: Operated by a Class P, 425 Ma, 60 hertz ballast. The ballast shall provide reliable lamp starting at 0 degrees F and shall be provided with a sound level rating of B or better. Lamp voltage variation shall not exceed plus five percent and minus 10 percent.

Radio interference filter shall be provided as indicated.

03. Fixture Wiring

a. Fixture wires shall be stranded tinned-copper construction, not smaller than No. 16 AWG. Insulation shall be silicone rubber type SF-2 and 200 degrees C rated. Conductor size, temperature rating, voltage rating and manufacturer shall be clearly marked on the insulation of each conductor.

b. Wires between lampholders and associated operating and starting equipment shall have the same ampacity rating as leads from the ballast. Wiring within the fixtures shall conform to the requirements of the NEC.

c. Wires shall be taped at all points of abrasion. No splices shall be permitted within fixtures other than as required to connect lampholders and ballasts. Wireways and wiring channels shall have rounded edges or bushed holes wherever conductors pass through. Insulated bushings shall be installed at points of entrance and exit of wiring.

d. Fixture Grounding. Unless otherwise specified, the housing of each ballasted lighting fixture shall be provided with a separate, factory-installed grounding device. The grounding device is to be used for connecting a separate grounding conductor to the fixture housing.

E. Fixture Hardware

01. Latch and release mechanism, hinges, pins and other retaining parts of fixtures; screws, bolts or other assembly and mounting parts shall be manufactured of Type 316 stainless steel. All springs shall be heavy duty stainless steel. All retaining hardware shall be self-retaining.

02. Light transmitting elements of the fixture shall be framed to permit replacement of panels in the frames without the use of tools other than screwdriver or pliers. Panels shall be held in the frames in a neat, rattle-free manner that will provide proper tolerance for normal expansion and contraction.

03. Internal brackets shall be fabricated from sheet steel, zinc coated after fabrication.

04. Gaskets, sealants and adhesives subjected to high temperature shall be formed from silicone rubber. Other gaskets shall be neoprene or as indicated.

05. Provide bolts, nuts, washers, screws, nails, rivets and other fastenings necessary for proper erection or assembly of work. When exposed to the atmosphere shall be made of 18-8 stainless steel. Fasteners within the housing shall be made of zinc plated, bright iridite, steel or electrogalvanized, gray. All nuts shall have captive externally footed lockwashers.

## 2.02 FIXTURE MOUNTING HARDWARE

A. General Requirements. The fixtures shall be provided with brackets, straps, canopies and stems, poles and miscellaneous hardware suitable for the mounting method specified.

B. Mounting brackets shall be secured to housing, quantity and spacing shall be as indicated. When exposed to public view, hardware shall be fabricated and finished in matching material to fixture body.

C. Canopies, holders and similar parts shall be drawn or spun in one piece with a minimum 0.026 inch finished thickness.

D. Tubine used for stems shall be seamless drawn with a minimum of 1/16 inch wall thickness of size and length as indicated. They shall be provided for all pendant mounted fixtures of length as required for the specified mounting height with swivel hangers or ball aligners as required.

*Insert required wind load rating.*

E. Light poles. Of the type, configuration and dimensions, indicated. The pole must resist wind loads of -B- mph with a maximum deflection of five percent when fully loaded by their own weight, weight and wind resistance of luminaires they support, and any externally applied loads. Furnish poles as indicated with four by six inch handhole with flushcover, luminaire mounting (tenon/bracket), base cover and all mounting hardware including anchor bolts, nuts, washers and baseplate to permit accurate alignment and installation of pole and luminaire as indicated.

## 2.03 LAMPS

A. General Requirements. Each lighting fixture shall be provided with the number, type, and wattage of lamps required by the contract drawings. All lamps used in the illumination system shall be of standard manufacture, readily available, and of the highest efficacy and life consistent with other requirements of the illumination system.

B. Incandescent Lamps: Rated for 120 volts and minimum 2000 hours life.

C. Fluorescent Lamps: Output as indicated with -B- color.

D. Mercury Lamps: Color, white deluxe. Reflector face, if required, shall be clear.

E. Metal Halide Lamps: Clear and provided with position-oriented mogul bases. Photometric characteristics shall provide lamp maximum luminous output while lamp operates in the horizontal position.

F. High Pressure Sodium Lamps: Clear and suitable for all operating positions.

*Delete the following equipment specification if in another section, i.e., Section 16610.*

2.04 AUXILIARY LIGHTING EQUIPMENT

A. General Requirements. Auxiliary lighting equipment intended to supply illumination in the event of failure of normal power supply shall conform to the applicable requirements of: UL924, NFPA-70 and NFPA-101.

B. Unitized battery packs mounted integral with fluorescent fixtures shall energize upon failure of normal power and shall provide approximately the constant light output delivered under normal power operation, for a period not less than 90 minutes. The unit shall contain a transistorized inverter ballast, a transfer relay and associated circuitry, a battery charger and batteries of nickel-cadmium. In addition, a test button and derangements signal light shall be provided to monitor the charging function.

C. Battery packs mounted remote from luminaires shall conform to the applicable requirements of UL 924. The battery-powered source provides continuous power to lighting loads, consisting of any mix of HID, fluorescent, or incandescent lamps. During short power interruptions, brownout conditions or a total lapse of normal ac power, it supplies the full rated load at both 120 and 277 volts for ninety minutes.

The unit consists of an inverter, battery charger, batteries, transfer circuitry, indicators and test switch housed in NEMA I enclosures.

A lockable hinged front door will permit ready access to all electronics and batteries. Batteries mounted on swing-out shelves permit full access for servicing without removing them from their installed location. Electronic equipment, mounted for easy access, utilizes plug-in printed circuit boards and modularized subassemblies to minimize down time and repair skills.

01. Inverter: Solid state, of the SCR type, supplies 60 Hz power at both 120 and 277V ac at its full rated load. The load may consist of any mix of HID, fluorescent, or incandescent lamps or other approved loads at either voltage, not damaged by overload and with positive protection against short circuits.

02. Battery Charger: Completely automatic, solid state, pulse type, dual rate device maintains the batteries in a fully-charged position. When required, a high rate charge current returns the batteries to 95 percent of charge capacity and the float rate charger completes the charge and maintains the batteries at their full capacity. The battery charger incorporates a capability to accommodate either lead acid or alkaline batteries.

03. Batteries: Long life lead acid batteries of the communication type meet or exceed the performance specifications. The 12-volt, 70-ampere hour batteries incorporating visual charge level indicators shall be connected in series parallel to provide a 48V dc supply. The nickel cadmium batteries shall be warranteed for at least 15 years.

04. Transfer System: Senses loss of normal ac power or brownout condition and transfers the inverter power source from rectified ac to battery power. All switching is solid state and accomplished in such a manner to provide continuous power to the load without any interruption.

05. Indicators and Switches: Provided to determine equipment status, charge mode, battery condition and system malfunction. They consist of a green ac power light, an amber derangement signal light (DSL) and a red light to indicate inverter operation. Battery charge status is indicated by a dc voltmeter and visual indicators on each battery. An external press to test switch simulates loss of normal ac power. An internal switch provides disconnect for unit servicing.

*Delete the following equipment specification if in another section, i.e., Section 16930.*

2.05 LIGHTING CONTROL EQUIPMENT

A. General Requirements. All lighting control components shall be suitable for the lighting system specified and compatible for interface with other associated control devices. Lighting control components shall be rated for continuous service and operate satisfactorily in every respect while the branch circuit power supply voltage to each system is within a 105 to 130 volt range at 60 hertz. Electrical contacts shall have precious metal surfaces.

B. Lighting Contractors:

   01. Conforms to the applicable requirements of UL 508.

   02. Electrically operated and mechanically held.

   03. Rated at 600 volts, 60 hertz with ampere rating, number of poles and enclosure as indicated.

C. Lighting Relays:

   01. Conforms to the applicable requirements of UL 508.

   02. Electrically operated and mechanically held.

   03. Rated at 600 volts, 60 hertz, 30 amperes with number of poles and enclosure as indicated.

D. Time Switches:

   01. Conforms to the applicable requirements of UL 887.

   02. Prewired with astronomic dial, 36-hour synchronous reserve power motor.

   03. Manual on-auto-off bypass switches for up to three individual circuits.

   04. Rated at 277 volts, 60 hertz, 40 amperes continuous duty with number of poles, throws and enclosure as rated.

E. Photoelectric Sensor:

   01. Conform to the applicable requirements of UL 773.

02. Operation in temperature range of minus 50 degrees C to plus 60 degrees C.

03. Dusk to dawn operation with adjustments from two to 50 footcandles with a five-second time delay to preclude false switching.

04. Weatherproof and tamperproof.

F. Light Intensity Controls:

01. Enclosed, continuously-adjustable, and completely solid state for the control voltage and rated load indicated.

02. Incandescent systems.

03. Fluorescent systems.

04. HID systems.

G. Wall Switches

01. Fed. Spec. W-S-896, types II and III shall apply. Switches installed in hazardous areas shall be the explosion-proof type in accordance with the NEC and as indicated on the drawings.

02. Switches shall be single unit, toggle, butt contact, quiet type with an integral mounting strap.

03. Wall switches for remote control shall be the momentary contact type suitable for mounting in a single gang outlet box space and compatible with the standard design wall plates.

04. Switch Ratings

    a. For 120 volt circuits: 20 amperes at 120 volts AC.

    b. For 277 volt circuits: 20 amperes at 277 volts AC.

05. Switches shall be connected to the wiring with screw clamp type terminals.

06. Wall Plates:

    a. Type 304 stainless steel.

b. Standard designs so the products of different manufacturers will be interchangeable.

c. Where switches are mounted adjacent to each other, the plates shall be common for each of the groups of switches.

07. Incorporate barriers between switches within multigang outlet boxes where required by the NEC.

## PART 3 - EXECUTION

3.01 LIGHTING FIXTURES. Lighting fixtures shall be installed in accordance with the manufacturer's instructions, complete with lamps, hangers, brackets, poles, fittings, and accessories, ready for operation as indicated.

A. Align, mount and level the lighting fixtures uniformly.

B. Avoid inteference with and provide clearance for equipment. Where the indicated locations for the lighting fixtures conflict with the locations for equipment, change the locations for the lighting fixtures by the minimum distances necessary as approved by the Engineer.

C. For suspended lighting fixtures, the mounting heights shall provide the clearances between the bottoms of the fixtures and the finished floors as indicated.

D. Lighting fixture supports shall provide support for all of the fixtures. Anchor supports to the structural slab or to structural members as indicated. Supports shall maintain the fixture positions after cleaning and relamping.

E. Surface mounted lighting fixtures shall be bracketed rigidly from the mounting surfaces. A 1/4-inch clearance between surfaces shall be provided when the fixture is flat mounted against concrete surfaces. Fixtures shall be installed with a non-cumulative dimensional alignment tolerance of 1/16 inch and when mounted in continuous runs shall be mounted with one inch spacing between individual fixtures. Nipples carrying wires between fixtures shall be water-tight.

F. Where aluminum is placed in contact with dissimilar materials, except galvanized steel, zinc or stainless steel, treat contact surfaces as follows:

01. When in contact with dissimilar metals apply a prime coat of zinc chromate primer followed by two coats of aluminum and masonry paint.

02. When in contact with concrete, masonry and plaster apply to aluminum contact surfaces zinc chromate primr, bituminous paint, aluminum metal and masonry paint or pressure tape.

03. When in contact with wood or other absorptive materials, apply two coats of aluminum house paint to such materials and protect aluminum contact surfaces with bituminous paint.

G. Welding:

01. Locate welds in assemblies to be anodized to conceal visible discoloration in the heat-affected zone.

02. Where weld metal will be exposed after anodizing, select filler alloys to closely match composition of base metal. Follow parent metal manufacturer's recommendations for such filler alloys.

H. Pendant fixtures shall be provided with swivel hangers to assure a plumb installation and have a minimum 25-degree swing from horizontal in all directions. Single unit suspended fluorescent fixtures shall have twin stem hangers. Multiple unit or continuous units shall have a tubing or stem for wiring at one point and tubing provided for each unit length of chassis including one at each end. Tubing shall not be less than 3/16 inch in diameter. Motion of swivels or hinged joints shall not cause sharp bends in conductors or damage to insulation. For heavy pendant mounted fixtures, where support independent of box is required and where conduit and outlet boxes are installed on surface, safety swivel hangers with fixture studs shall be provided.

I. Fixtures to be pole mounted shall be installed in accordance with the manufacturer's recommended installation practices as indicated.

J. Required lamps shall be provided in each lighting fixture as soon as fixtures are properly installed.

3.02 BALLASTS. Ballasts, other than those mounted integrally within luminaries, shall be installed as indicated, and in such a manner that the ballast is protected from weather, moisture, and other atmospheric conditions, and in such a manner that the ambient temperature surrounding the ballast will not cause the temperature of the ballast housing hot spot to exceed UL requirements. Voltage drop to lamp, due to remote mounting shall not exceed one percent of the nominal lamp voltage. Secondary ballast conductors shall have 1000 volt insulation. When more than one ballast is mounted at one location, the minimum spacing between ballasts shall be 6 inches in a horizontal direction and 12 inches in a vertical direction.

3.03    LIGHT POLES. Installation of light poles shall be in accordance with the manufacturer's recommended installation practices as indicated.

3.04    CONCRETE BASES: Obtain necessary templates and anchor kits before starting work.

*Delete the following equipment specification if in another section, i.e., Section 1660.*

3.05    AUXILIARY LIGHTING EQUIPMENT:

    A. Install as indicated and in accordance with manufacturer's instructions.

    B. Anchor firmly in place.

    C. Test and adjust for proper operation in accordance with the manufacturer's instructions.

*Delete the following equipment specification if in another section, i.e., Section 16930.*

3.06    LIGHTING CONTROL DEVICES. Lighting control devices shall be installed in accordance with the manufacturer's recommended installation practices as indicated.

    A. Where indicated incorporate the components in panelboards behind separate doors and mount them on sound absorbing materials.

    B. Install circuit breaker or fuse protection for the control circuits.

    C. Mount the switches on the striker plate side of the doors.

3.07    FIELD QUALITY CONTROL AND INSPECTION:

    A. Luminaries, lamps and associated hardware shall be inspected prior to and after installation to assure that they are of the quality and type as specified herein and as shown on the luminaire pallette, and are free of defects and damage.

B. Luminaires and lighting equipment shall be delivered to the project site complete, with suspension accessories, canopies, hickeys, castings, sockets, holders, reflectors, ballasts, diffusing materials, louvers, frames, recessing boxes, and related items, completely wired and assembled.

C. Whenever practicable, lighting systems shall be tested at the same time that the distribution panelboard or switchboard is tested.

# Glossary

*Accent Lighting*: Directional lighting to emphasize a particular object or to draw attention to a part of the field of view.

*Absorption*: The dissipation of light within a surface or medium.

*Accommodation*: The process by which the eye changes focus from one distance to another.

*Adaptation*: The process by which the visual system becomes accustomed to more or less light than it was exposed to during an immediately preceding period. It results in a change in the sensitivity of the eye to light.

*Air Fitting*: (Air bonnet, Air hood, Air saddle, Air box) A fitting which is mounted to an air handling luminaire and connects to the primary air duct by flexible ducting. It normally contains one or two volume controls.

*Alternating Current (AC)*: Flow of electricity which cycles or alternates direction many times per second. The number of cycles per second is referred to as frequency. Most common frequency used in this country is 60 Hertz (cycles per second).

*Ambient Lighting*: General lighting, or lighting of the surround (as opposed to task lighting or the lighting of the object one is looking at). It can be produced by direct lighting from recessed surface or stem mounted luminaires, or by indirect lighting which is wall or stem mounted, built into furniture or free standing.

*Amperes* (amps or A): The unit of measurement of electric current.

*Baffle*: An opaque or translucent element that serves to shield a light source from direct view at certain angles, or serves to absorb unwanted light.

*Ballast*: An auxiliary device consisting of induction windings wound around a metal core and sometimes including a capacitor for power correction. It is used with fluorescent and HID lamps to provide the necessary starting voltage and to limit the current during operation.

*"Batwing" Distribution*: Candlepower distribution which serves to reduce glare and veil reflections by having its maximum output in the $30^o$ to the $60^o$ zone from the vertical and with a candlepower at nadir (0 degrees) being 65% or less than maximum candlepower. The shape is similar to a bat's wing.

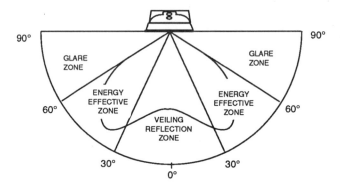

**Batwing Distribution**

In fluorescent luminaires the batwing distribution is generally found only in the plane perpendicular to the lamps..

*Beam Spread*: The angle enclosed by two lines which intersect the candlepower distribution curve at the points where the candlepower is equal to ten percent of its maximum.

*Branch Circuit*: An electrical circuit running from a service panel having its own overload protection device.

*Brightness (Luminance)*: The degree of apparent lightness of a surface; its brilliancy; concentration of candlepower. Brightness is produced by either a self-luminous object, by light energy transmitted through objects or by reflection. Unit of measurement of brightness is the footlambert (fl).

*"BX" Cable*: A cable comprised of a flexible metallic covering inside of which are two or more insulated wires for carrying electricity.

*Candela:* The unit of measurement of luminous intensity of a light source in a given direction.

*Candlepower:* Luminous intensity expressed in candelas.

*Candlepower Distribution Curve:* A graphic presentation of the distribution of light intensity in a given plane of a lamp or luminaire. It is determined by photometric tests. The curve is generally polar, representing the variation of luminous intensity of a lamp or luminaire in a plane through the light center.

*Capacitor:* An electric energy storage device which when built into or wired to a ballast changes it from low to high power factor.

*Cavity Ratio:* A number indicating cavity proportions calculated from length, width and height.

*Ceiling Cavity Ratio:* A numerical relationship of the vertical distance between luminaire mounting height and ceiling height to room width and length. It is used with the Zonal Cavity method of calculating average illumination levels.

*Circuit Breaker:* Resettable safety device to prevent excess current flow.

*Class "P" Ballast:* Contains a thermal protective device which deactivates the ballast when the case reaches a certain critical temperature. The device resets automatically when the case temperature drops to a lower temperature.

*Coefficient of Utilization (CU):* A ratio representing the portion of light emitted by a luminaire in any particular installation that actually gets down to the work plane. The coefficient of utilization thus indicates the combined efficiency of the luminaire, room proportions and room finish reflectances. The ratio of the luminous flux (lumens) from a luminaire is calculated as received on the work-plane to the luminous flux emitted by the luminaire's lamps alone.

*Cold Cathode Lamp:* An electric-discharge lamp whose mode of operation is that of a glow discharge.

*Color Rendering Index (CRI):* Measure of the degree of color shift objects undergo when illuminated by the light source as compared with the color of those same objects when illuminated by a reference source of comparable color temperature.

*Color Temperature:* The absolute temperature of a blackbody radiator having a chromaticity equal to that of the light source.

*Cone Reflector*: Parabolic reflector that directs light downward thereby eliminating brightness at high angles.

*Contrast*: The difference in brightness (luminance) of an object and its background.

*Contrast Rendition Factor (CRF)*: The ratio of visual task contrast with a given lighting environment to the contrast with sphere illumination. Contrast measured under sphere illumination is defined as 1.00.

*Cool Beam Lamps*: Incandescent PAR lamps that use a special coating (dichronic interference filter) on the reflectorized portion of the bulb to allow heat to pass out the back while reflecting only visible energy to the task, thereby providing a "cool beam" of light.

*Cut-off Luminaires*: Outdoor luminaires that restrict all light output to below $85°$ from vertical.

*Cut-off Angle* (of a luminaire): The angle from the vertical at which a reflector, louver, or other shielding device cuts off direct visibility of a light source. It is the complementary angle of the shielding angle. In the case of reflector-type lightshields it is also important to ascertain the cut-off angle to the reflected image of the light source as this is often almost as bright as the source itself.

*Dimming Ballast*: Special fluorescent lamp ballast, which when used with a dimmer control, permits varying light output.

*Direct Current (DC)*: Flow of electricity continuously in one direction from positive to negative.

*Direct Glare*: Glare resulting from high luminances or insufficiently shielded light sources in the field of view. It usually is associated with bright areas, such as luminaires, ceilings and windows which are outside the visual task or region being viewed.

*Discharge Lamp*: A lamp in which light (or radiant energy near the visible spectrum) is produced by the passage of an electric current through a vapor or a gas.

*Discomfort Glare*: Glare producing discomfort. It does not necessarily interfere with visual performance or visibility.

*Distribution Panel*: Box containing circuit breakers or fuses where power is distributed to branch circuits.

*Efficacy*: See Lamp Efficacy.

*Efficiency*: See Luminaire Efficiency.

*Equivalent Sphere Illumination (ESI):* The level of sphere illumination which would produce task visibility equivalent to that produced by a specific lighting environment.

Suppose a task at a given location and direction of view within a specific lightings system has 100 fc of illumination.

Suppose this same task is now viewed under sphere lighting and the sphere lighting level is adjusted so that the task visibility is the same under the sphere lighting as it was under the lighting system. Suppose the lighting level at the task from the sphere lighting is 50 fc for equal visibility. Then the Equivalent Sphere Illumination of the task under the lighting system would be 50 ESI fc.

*"ER" (Elliptical Reflector)*: Lamp whose reflector focuses the light about 2" ahead of the bulb, reducing light loss when used in deep baffle downlights.

*Extended Life Lamps*: Incandescent lamps that have an average rated life of 2500 or more hours and reduced light output compared to standard general service lamps of the same wattage.

*Floodlighting*: A system designed for lighting a scene or object to a luminance greater than its surroundings. It may be for utility, advertising or decorative purposes.

*Fluorescent Lamp*: A low-pressure mercury electric-discharge lamp in which a fluorescing coating (phosphor) transforms some of the ultraviolet energy generated by the discharge into light.

*Flux*: Continuous flow of luminous energy.

*Footcandle (fc)*: The unit of illuminance when the foot is taken as the unit of length. It is the illuminance on a surface one square foot in area on which there is a uniformly distributed flux of one lumen.

*(Raw) Footcandles*: Same as footcandles. This term is sometimes used in order to differentiate between ordinary footcandles and ESI footcandles. (Footcandles or Raw Footcandles refer only to the quantity of illumination. ESI footcandles refer to task visibility by considering both the quantity and quality of illumination.)

*Foot Lambert (fl)*: A unit of luminance of a perfectly diffusing surface emitting or reflecting light at the rate of one lumen per square foot.

*Fuse*: Replaceable safety device to prevent excess current flow.

*General Lighting*: See Ambient Lighting.

*General Service Lamps*: "A" or "PS" incandescent lamps.

*Glare*: The sensation produced by luminance within the visual field that is sufficiently greater than the luminance to which the eyes are adapted to cause annoyance, discomfort, or loss in visual performance and visibility.

*Direct Glare*: Glare resulting from high brightness or insufficiently shielded light sources in the field of view.

*Reflected Glare*: Glare resulting from specular reflections of high brightness sources in polished surfaces in the field of view. Also see Veiling Reflections.

*Greenfield*: Flexible metallic tubing for the protective enclosure of electric wires.

*Grounding*: Connection of electric components to earth for safety.

*Ground Relamping*: Relamping of a group of luminaires at one time to reduce relamping labor costs.

*Heat Extraction*: The process of removing heat from a luminaire by passing return air through the lamp cavity.

*High Intensity Discharge (HID) Lamp*: A discharge lamp in which the light producing arc is stabilized by wall temperature, and the arc tube has a bulb wall loading in excess of three watts per square centimeter. HID lamps include groups of lamps known as mercury, metal halide, and high pressure sodium.

*High Output Fluorescent Lamp*: Operates at 800 or more milliamperes for higher light output than standard fluorescent lamp (430MA).

*High Pressure Sodium (HPS) Lamp*: High intensity discharge (HID) lamp in which light is produced by radiation from sodium vapor. Includes clear and diffuse-coated lamps.

*Incandescent Lamp*: A lamp in which light is produced by a filament heated to incandescence by an electric current.

*Instant Start Fluorescent Lamp*: A fluorescent lamp designed for starting by a high voltage without preheating of the electrodes.

*Inverse Square Law*: The law stating that the illuminance at a point on a surface varies directly with the intensity of a point source, and inversely as the square of the distance between the source and the point. If the surface at the point is normal to the direction of the incident light, the law is expressed by $fc = cp/d^2$.

*Isolux Chart*: A series of lines plotted on any appropriate set of coordinates, each line connecting all the points on a surface having the same illumination.

*Junction Box*: A metal box in which circuit wiring is spliced. It may also be used for mounting luminaires, switches or recepticles.

*Kilowatt-Hour (KWH)*: Unit of electrical power consumed over a period of time. KWH=watts/1000xhours used.

*Lamp*: An artificial source of light (also a portable luminaire equipped with a cord and plug).

*Lamp Efficacy*: The ratio of lumens produced by a lamp to the watts consumed, expressed as lumens per watt (LPW).

*Lamp Lumen Depreciation (LLD)*: Multiplier factor in illumination calculations for reduction in the light output of a lamp over a period of time.

*Light*: Radiant energy that is capable of exciting the retina and producing a visual sensation. The visible portion of the electromagnetic spectrum extends from about 380 to 770 nm.

*Light Loss Factor (LLF)*: A factor used in calculating the level of illumination that takes into account such factors as dirt accumulation on luminaire and room surfaces, lamp depreciation, maintenance procedures and atmosphere conditions. See Maintenance Factor.

*Light Output:* Amount of light produced by a light source such as a lamp. The unit most commonly used to measure light output is the lumen.

*Lens*: Used in luminaires to redirect light into useful zones.

*Local Lighting*: Lighting designed to provide illuminance over a relatively small area or confined space without providing any significant general surrounding lighting.

*Louver*: A series of baffles used to shield a source from view at certain angles or to absorb unwanted light. The baffles usually are arranged in a geometric pattern.

*Long Life Lamps*: See Extended Life Lamps.

*Low Pressure Sodium Lamp*: A discharge lamp in which light is produced by radiation of sodium vapor at low pressure producing a single wavelength of visible energy, i.e. yellow.

*Low Voltage Lamps*: Incandescent lamps that operate at 6 to 12 volts.

*Lumen*: The unit of luminous flux. It is the luminous flux emitted within a unit solid angle (one steradian) by a point source having a uniform luminous intensity of one candela.

*Luminaire*: A complete lighting unit consisting of a lamp or lamps together with the parts designed to distribute the light, to position and protect the lamps and to connect the lamps to the power supply.

*Luminaire Dirt Depreciation (LDD)*: The multiplier to be used in illuminance calculations to relate the initial illuminance provided by clean, new luminaires to the reduced illuminance that they will provide due to dirt collection on the luminaires at the time at which it is anticipated that cleaning procedures will be instituted.

*Luminaire Efficiency:* The ratio of luminous flux (lumens) emitted by a luminaire to that emitted by the lamp or lamps used therein.

*Luminance*: The amount of light reflected or transmitted by an object.

*Lux*: The metric unit of illuminance. One lux is one lumen per square meter (lm/m2).

*Maintenance Factor (MF)*: A factor used in calculating illuminance after a given period of time and under given conditions. It takes into account temperature and voltage variations, dirt accumulation on luminaire and room surfaces, lamp depreciation, maintenance procedures and atmosphere conditions.

*Matte Surface*: A non-glossy dull surface, as opposed to a shiny (specular) surface. Light reflected from a matte surface is diffuse.

*Mercury Lamp*: A high intensity discharge (HID) lamp in which the major portion of the light is produced by radiation from mercury. Includes clear, phosphor-coated and self-ballasted lamps.

*Metal Halide Lamp*: A high intensity discharge (HID) lamp in which the major portion of the light is produced by radiation from mercury. Includes clear, phosphor-coated and self-ballasted lamps.

*Nadir*: Vertically downward directly below the luminaire or lamp; designated as $0°$.

*Outlet Box*: See Junction Box.

*"PAR" Lamps*: Parabolic aluminized reflector lamps which offer excellent beam control, come in a variety of beam patterns from very narrow spot to wide flood, and can be used outdoors unprotected because they are made of "hard" glass that can withstand adverse weather.

*Parabolic Louvers*: A grid of baffles which redirects light downward and provides very low luminaire brightness.

*Pattern Control*: A blade, in the air passage of an air handling luminaire, which sets the direction of air flow from the luminaire.

*Plug-in Wiring*: Electrical distribution system which has quick-connect wiring connectors.

*Point Method Lighting Calculation*: A lighting design procedure for predetermining the illuminance at various locations in lighting installations, by use of luminaire photometric data.

*Polarization:* The process by which the transverse vibrations of light waves are oriented in a specific plane. Polarization may be obtained by using either transmitting or reflecting media.

*Power Factor*:   Ratio of :   $\dfrac{\text{watts}}{\text{volts} \times \text{amperes}}$

Power factor in lighting is primarily applicable to ballasts. Since volts and watts are usually fixed, amperes (or current) will go up as power factor goes down. This necessitates the use of larger wire sizes to carry the increased amount of current needed with Low power Factor (L.P.F.) ballasts. The addition of a capacitor to a L.P.F. ballast converts it to a H.P.F. ballast.

*Preheat Fluorescent Lamp:* A fluorescent lamp designed for operation in a circuit requiring a manual or automatic starting switch to preheat the electrodes in order to start the arc.

*"R" Lamps*: Reflectorized lamps available in spot (clear face) and flood (frosted face).

*Rapid Start Fluorescent Lamp*: A fluorescent lamp designed for operation with a ballast that provides a low-voltage winding for preheating the electrodes and initiating the arc without a starting switch or the application of high voltage.

*Raw Footcandles*: See Footcandles.

*Reflection*: Light striking a surface is either absorbed, transmitted, or reflected. Reflected light is that which bounces off the surface, and it can be classified as specular or diffuse reflection. Specular reflection is characterized by light rays which strike and leave a surface at equal angles. Diffuse reflection leaves a surface in all directions.

*Reflectance*: Sometimes called reflectance factor. The ratio of reflected light to incident light (light falling on a surface). Reflectance is generally expressed in percent.

*Reflected Glare*: Glare resulting from specular reflections of high luminances in polished or glossy surfaces in the field of view. It usually is associated with reflections from within a visual task or areas in close proximity to the region being viewed.

*Refraction*: The process by which the direction of a ray of light changes as it passes obliquely from one medium to another in which its speed is different.

*Romex*: A cable comprised of flexible plastic sheathing inside of which are two or more insulated wires for carrying electricity.

*Room Cavity Ratio (RCR)*: A numerical relationship of the vertical distance between work plane height and luminaire mounting height to room width and lengh. It is used with the Zonal Cavity method of calculating average illumination levels.

*Rough Service Lamps*: Incandescent lamps designed with extra filament supports to withstand bumps, shocks and vibrations with some loss in lumen output.

*Self-ballasted Mercury Lamps*: Any mercury lamp of which the current-limiting device is an integral part.

*Service Entrance*: Point at which power utility wires enter a building.

*Shielding:* An arrangement of light-controlling material to prevent direct view of the light source.

*Shielding Angle* (of a luminaire): The angle from the horizontal at which a light source first becomes visible. It is the complementary angle of the cut-off angle. In the case of a luminaire shielded by a reflector or parabolic cell louver, it is important to ascertain also the shielding angle to the reflected image of the light source, as this is often almost as bright as the source itself.

*Silvered Bowl Lamps*: Incandescent 'A' lamps with a silver finish inside the bowl portion of the bulb. Used for indirect lighting.

*Sound Transmission* (room to room): Sound passing from one room to another, normally through an air return plenum. Also called Crosstalk.

*Sound Transmission Class (STC):* A number rating system for room to room sound transmission through air handling luminaires. The higher the STC number, the lower the level of sound transmission.

*Spacing Ratio (SR):* The ratio of the distance between luminaire centers to the height above the work plane. The maximum spacing ratio for a particular luminaire is determined from the candlepower distribution curve for that luminaire and, when multiplied by the mounting height above the work plane, gives the maximum spacing of lunimaires at which even illumination will be provided.

*Spectral Energy Distribution (SED)* Curves: A plot of the level of energy at each wavelength of a light source.

*Speech Privacy*: The extent to which people in public areas can speak without being overheard by others or disturbing others.

*Sphere Illumination*: The illumination on a task from a source providing equal luminance in all directions about that task, such as an illuminated sphere with the task located at the center.

*Task*: That which is to be seen. The visual function to be performed.

*Task Lighting*: Lighting directed to a specific surface or area that provides illumination for visual tasks.

*Three-way Lamps*: Incandescent lamps that have two separately switched filaments permitting a choice of three levels of light such as

30/70/100, 50/100/150 or 100/200/300 watts. They can only be used in the base down position.

*Transformer:* A device to raise or lower electric voltage.

*Transmission*: The passage of light through a material.

*Tungsten-Halogen Lamp*: A gas filled tungsten incandescent lamp containing a certain proportion of halogens.

*Veiling Reflections*: The reflections of light sources in the task which reduce the contrast between detail and background (e.g. between print and paper) thus imposing a "veil" and decreasing task visibility. (Veiling reflections are sometimes referred to as Reflected Glare but the latter term is properly used only when specular reflections of the light source in the task and background are so bright as to be disturbing, whereas veiling reflections are often much less obvious. Their subtle effect in reducing contrast and thus visibility is nonetheless present.)

*Vibration Service Lamps*: See Rough Service Lamps.

*Visual Comfort Probability (VCP)*: The rating of a lighting system expressed as a percent of people who, when viewing from a specified location and in a specified direction, will be expected to find it acceptable in terms of discomfort glare.

*Visual Edge*: The line on a isolux chart which has a value equal to 10% of the maximum illumination.

Visual Field: The field of view that can be perceived when the head and eyes are kept fixed.

Visual Comfort Portability (VCP): A discomfort glare calculation that predicts the percent of observers positioned at a specific location, (usually four feet in front of the center of the rear wall), who would be expected to judge a lighting condition to be comfortable. VCP rates the luminaire in its environment, taking into account such factors as illumination level, room dimensions and reflectances, luminaire type, size and light distribution, number and location of luminaires, and observer location and location and line of sight. The higher the VCP, the more comfortable the lighting environment. IES has established a value of 70 as the minimum acceptable VCP.

*Volt (V):* The unit for measuring electric potential. It defines the force or pressure of electricity.

*Wall Wash Lighting*: A smooth even distribution of light over a wall.

*Watt (W):* The unit for measuring electric power. It defines the power or energy consumed by an electrical device. The cost of operating an electrical device is determined by the watts it consumes times the hours of use. It is related to volts and amps by the following formula: Watts = Volts x Amps.

*Work Plane:* The plane at which work is done, and at which illumination is specified and measured. Unless otherwise indicated, this is assumed to be a horizontal plane 30 inches above the floor.

*Zonal Cavity Method Lighting Calculation*: A lighting design procedure used for predetermining the relation between the number and types of lamps or luminaires, the room characteristics, and the average illuminance on the work-plane. It takes into account both direct and reflected flux.

# Bibliography

Birren, Faber. *Light, Color and Environment*. New York: Van Nostrand-Reinhold, 1969.

Boyce, P.R. *Human Factors in Lighting*. New York: Macmillan, 1981.

Flynn, J.E., Mills, S.M. *Architectural Lighting Graphics*. New York: Van Nostrand-Reinhold, 1962.

Kaufman, John E. *IES Lighting Handbook*, 1981, Application Volume. New York: Illuminating Engineering Society.

Kaufman, John E.. *IES Lighting Handbook*, 1981, Reference Volume. New York: Illuminating Engineering Society

Lam, William M.C. *Perception and Lighting as Formgivers for Architecture*. New York: McGraw-Hill Book Co., 1977.

*Lessons in Lighting*. Jersey City, N.J.: Lightolier Inc., 1982.

McGuinness, W.J., Stein, B., Reynolds, J.S *Mechanical and Electrical Equipment for Buildings*. New York: John Wiley & Sons, 1980.

# LIST OF PLATES

## CHAPTER 1

FIGURE

| | |
|---|---|
| 1-1 | Spectrum |
| 1-2 | Footcandle Chart |
| 1-3 | Chart of Vision |
| 1-4 | Eye Chart *(IES)* |
| 1-5 | The Nearer An Object |
| 1-6 | More Contrast |
| 1-7 | Luminence—Reflect Light |
| 1-8 | Inverse Square Law |
| 1-9 | Candle Power Distribution Curve *(L)* |
| 1-10 | Glare in the Field *(L)* |
| 1-11 | Veiled Reflections |
| 1-12 | Task Visability |
| 1-13 | Location Viewing Direction *(L)* |
| 1-14 | Color Recognition |
| 1-15 | Wavelength *(IES)* |
| 1-16 | Black Body Locus *(IES)* |
| 1-17 | Perceived Color Effects from Lamps *(IES)* |
| 1-18 | Types of Light Modification |
| 1-19 | Transmission |
| 1-20 | Refraction *(IES)* |
| 1-21 | Reflections |
| 1-22 | Direct |
| 1-23 | Semi-Direct |
| 1-24 | General Diffuse |
| 1-25 | Direct-Indirect |
| 1-26 | Semi-Indirect |
| 1-27 | Indirect |
| 1-28 | ITL Report—Incandescent Downlight *(L)* |
| 1-29 | ITL Report—Flourescent Troffer *(L)* |
| 1-30 | Single/Multiple Luminaires Calculations *(L)* |
| 1-31 | Energy Effective Zone *(IES)* |
| 1-32 | 1LT.40W   HVP—Performance Spec. *(IES)* |
| 1-33 | Light Control Devices *(IES)* |

## CHAPTER 2

| | |
|---|---|
| 2-1 | Effect of Light on Interior Design |
| 2-2 | The Eye is Drawn to Brightest Surface in the Room |
| 2-3 | Glitter & Sparkle |
| 2-4 | Light & Shadow |
| 2-5 | Modeling *(Marlene Lee)* |
| 2-6 | Spacial, Function & Psychological Criteria *(L)* |
| 2-7 | Task Lighting *(Herman Miller)* |
| 2-8 | Accent Lighting *(Herman Miller)* |
| 2-9 | Ambient Lighting *(Herman Miller)* |
| 2-10 | Texture |
| 2-11 | Track Lighting |
| 2-12 | Wall Washer Data *(IES)* |
| 2-13 | Trough Lighting Chart *(IES)* |
| 2-14 | (not shown) |
| 2-15 | Maximum Beam Coverage *(L)* |
| 2-16 | Illuminance Values *(IES)* |
| 2-17 | Locating Light |
| 2-18 | Batwing |
| 2-19 | Photometric Curve *(L)* |

## CHAPTER 3

| | |
|---|---|
| 3-1 | Point Source Illumination |
| 3-2 | Single Incandescent Luminaire Data *(L)* |
| 3-3 | Single Incandescent Luminaire Data *(L)* |
| 3-4 | Lighting Performance Data *(L)* |
| 3-5 | Zonal Cavity Method |
| 3-6 | Light is Absorbed |
| 3-7 | CU Table 4LT.40W Luminaire *(L)* |
| 3-8 | Surface Dirt Depreciation *(L)* |
| 3-9 | Luminaire Dirt Depreciation *(L)* |
| 3-10 | Quick Calculator Chart *(L)* |
| 3-11 | Cavity Ratios Chart—Table I *(IES)* |
| 3-12 | Cavity Ratios Chart—Table II *(IES)* |
| 3-13 | Cavity Ratios Chart—Table III *(IES)* |
| 3-14 | Illustration for Lumen Point *(Lighting Technologies)* |
| 3-15 | Illustration for Lumen Point *(Lighting Technologies)* |
| 3-16 | Illustration of Worksheet *(Lighting Technologies)* |

3-17   B/W Perspective *(Lighting Technologies)*
3-18   Perspective From Targa *(Lighting Technologies)*

## CHAPTER 4

4-1    Continuous Spectrum *(IES)*
4-2    Line Spectrum
4-3    Lamp Selection Chart *(IES)*
4-4    Incandescent Filament Lamp
4-5    Base Shape Chart *(IES)*
4-6    Bulb Shape Chart *(IES)*
4-7    Light Source Efficiency *(IES)*
4-8    Mercury Lamps Spectral Power Distribution *(IES)*
4-9    Fluorescent Lamp *(IES)*
4-10   Spectral Distribution Curves *(IES)*
4-11   Types of Fluorescent Lamps *(IES)*
4-12   400W Phosphor-Coated Mercury Lamp *(IES)*
4-13   HID Lamps *(IES)*
4-14   High Pressure Sodium Lamps *(IES)*
4-15   Typical HID Lamp Bulb Sizes *(IES)*
4-16   Performance Characteristics of HID Lamps
4-17   Fluorescent Ballast *(IES)*
4-18   HID Quick Calculator Chart *(IES)*

## CHAPTER 5

5-1    Symbol List
5-2    Abbreviations
5-3    Single Line Diagram Symbols
5-4    Luminaire Schedule *(Marlene Lee)*
5-5    Lighting Calculations *(Marlene Lee)*
5-6    Example *(Marlene Lee)*
5-7    Example *(Marlene Lee)*
5-8    Example *(Marlene Lee)*
5-9    Example *(Marlene Lee)*
5-10   Example *(Marlene Lee)*
5-11   Example *(Marlene Lee)*
5-12   Example *(Marlene Lee)*
5-13   Example *(Marlene Lee)*
5-14   Example *(Marlene Lee)*
5-15   Example *(Marlene Lee)*

5-16    Example *(Marlene Lee)*
5-17    Example *(Marlene Lee)*

---

*IES* = Courtesy Illuminating Engineering Society

*L* = Courtesy Lightolier, Inc.

*Marlene Lee* = Marlene Lee Lighting Design

*Herman Miller* = Herman Miller Corp.

*Lighting Technologies* = Lighting Technologies, Inc.

All others by Frederic Jones

---